MW01435115

Teaching, Learning and Scaffolding in CLIL Science Classrooms

Benjamins Current Topics

ISSN 1874-0081

Special issues of established journals tend to circulate within the orbit of the subscribers of those journals. For the Benjamins Current Topics series a number of special issues of various journals have been selected containing salient topics of research with the aim of finding new audiences for topically interesting material, bringing such material to a wider readership in book format.

For an overview of all books published in this series, please see
benjamins.com/catalog/bct

Volume 115

Teaching, Learning and Scaffolding in CLIL Science Classrooms
Edited by Yuen Yi Lo and Angel M.Y. Lin
These materials were previously published in *Journal of Immersion and Content-Based Language Education* 7:2 (2019).

Teaching, Learning and Scaffolding in CLIL Science Classrooms

Edited by

Yuen Yi Lo
The University of Hong Kong

Angel M.Y. Lin
Simon Fraser University

John Benjamins Publishing Company
Amsterdam / Philadelphia

∞™ The paper used in this publication meets the minimum requirements of the American National Standard for Information Sciences – Permanence of Paper for Printed Library Materials, ANSI z39.48-1984.

DOI 10.1075/bct.115

Cataloging-in-Publication Data available from Library of Congress:
LCCN 2021009591 (PRINT) / 2021009592 (E-BOOK)

ISBN 978 90 272 0888 0 (HB)
ISBN 978 90 272 5979 0 (E-BOOK)

© 2021 – John Benjamins B.V.
No part of this book may be reproduced in any form, by print, photoprint, microfilm, or any other means, without written permission from the publisher.

John Benjamins Publishing Company · https://benjamins.com

Table of contents

Introduction

Teaching, learning and scaffolding in CLIL science classrooms 　　1
   Yuen Yi Lo and Angel M.Y. Lin

Chapters

Language focused episodes by monolingual teachers in English 　　17
Medium Instruction science lessons
   Jiangshan An, Ernesto Macaro and Ann Childs

The positioning of Japanese in a secondary CLIL science classroom in 　　43
Australia: Language use and the learning of content
   Marianne Turner

Teacher language awareness and scaffolded interaction in CLIL 　　63
science classrooms
   Daozhi Xu and Gary James Harfitt

Supporting students' content learning in Biology through teachers' use 　　85
of classroom talk drawing on concept sketches
   Caroline Ho, June Kwai Yeok Wong and Natasha Anne Rappa

Co-developing science literacy and foreign language literacy through 　　115
"Concept + Language Mapping"
   Peichang He and Angel M.Y. Lin

Scaffolding for cognitive and linguistic challenges in CLIL science 　　143
assessments
   Yuen Yi Lo, Wai-mei Lui and Mona Wong

Commentary

The role of language in scaffolding content & language integration in 　　169
CLIL science classrooms
   Kok-Sing Tang

Index　　182

Teaching, learning and scaffolding in CLIL science classrooms

Yuen Yi Lo and Angel M.Y. Lin
The University of Hong Kong | Simon Fraser University

This volume presents a collection of empirical studies examining the teaching and learning processes in science classrooms in Content and Language Integrated Learning (CLIL) contexts. It is a timely contribution to the rapidly growing body of CLIL research in response to scholars' consistent calls for more classroom-based research on the issues in integration of content and language teaching in lessons (Cenoz, Genesee, & Gorter, 2014; Nikula, Dalton-Puffer, Llinares, & Lorenzo, 2016). In this introduction, we will outline the key constructs in this volume, namely CLIL, science learning and scaffolding, in order to help readers understand the central theme of this volume and the connection of its various chapters.

1. Conceptualising Content and Language Integrated Learning (CLIL)

Since the term "CLIL" was coined in Europe in the 1990s, there have been numerous attempts to define the term. For example, Marsh (2002) explains CLIL "as a generic umbrella term which would encompass any activity in which a foreign language is used as a tool in the learning of a non-language subject in which both language and the subject have a joint curricular role," (p. 58) whereas Coyle, Hood and Marsh (2010) define CLIL as "a dual-focused educational approach in which an additional language is used for the learning and teaching of both content and language" (p. 1). These definitions highlight some key characteristics of CLIL, including learning of non-language content subject (e.g., science) through a foreign or additional language (which we will refer to as an L2 hereafter), as well as the important role played by both content and language. However, from these two definitions also emerge some differences or ambiguity – Does CLIL include "any activity" or does it refer to an "educational approach"? What does "joint curricular role" or "dual-focused approach" imply, regarding the balance between content teaching and language teaching in the curriculum and instruction? Theoretically speaking,

https://doi.org/10.1075/bct.115.01lo
© 2021 John Benjamins Publishing Company

the term "content and language integration" only highlights the core principle of integrating content and language learning, which is based on the inextricable relationship between content and language (Lemke, 1990), and the assumed benefits for L2 learning as proposed by various second language acquisition theories (see the edited volume by Llinares and Morton [2017] for a more detailed discussion). However, this general principle or theoretical position does not specify how such an integration can or should be achieved at the different levels of curriculum selection and sequencing, activities design and pedagogical strategies. Hence, different researchers may interpret "integration" at different levels, including teaching activities, pedagogical approaches, curriculum designs and educational programmes. That may explain why Cenoz et al. (2014) conclude, "the scope of CLIL is not clear-cut and, as a consequence, its core features cannot be clearly identified" (p. 247).

Perhaps with a view to clarifying the meaning of CLIL, there have been some attempts to distinguish CLIL from other content-based L2 education programmes, such as English as the Medium of Instruction (EMI), immersion programmes and content-based instruction. Comparison has been made in such aspects as goals, student population, teacher profile, the target language and the balance between content and language (e.g., Coyle et al., 2010; Lasagabaster & Sierra, 2010). Yet, these attempts have been challenged, mainly because the implementation of CLIL itself (and also the implementation of other programmes) varies so much that every claim about the unique features of CLIL can be refuted by some examples in some contexts (Somers & Surmont, 2012).

Another approach to define CLIL and other programmes is to put them along a continuum of content and language integration (e.g., Lin, 2016; Macaro, 2018). One end of the continuum features those more "content-driven" programmes (e.g., immersion programmes), whereas on the other end are those "language-driven" programmes (e.g., the conventional, often isolated L2 learning lessons). The merit of such a continuum is to acknowledge the fact that content and language are always integrated. It is the degree of such integration and which aspect the programmes put more emphasis on that differentiate the programmes. Following this line of differentiation, CLIL could be placed somewhere in the middle of the continuum, as its name suggests more or less equal emphasis on both content and language aspects. However, we would argue that this only represents the "ideal" implementation of CLIL, since research evidence has hardly demonstrated balanced instruction or integration of content and language in actual practice (Dalton-Puffer, 2007).

In recent years, the consensus in the field seems to adopt CLIL as an umbrella term which encompasses different kinds of content-based programmes. The advantage of doing so is to enable researchers and educators to learn from the experience and research evidence from different educational settings (Cenoz et al., 2014; Dalton-Puffer, Llinares, Lorenzo, & Nikula, 2014; Lin, 2016). However, adopting this approach does not mean different programmes are treated as the same, because they are obviously not. Instead, researchers are expected to describe their research contexts in greater detail, so that their counterparts in other contexts can critically examine to what extent the findings are applicable.

Following this approach, this volume adopts CLIL as an umbrella term, and it includes studies contextualised in different educational settings and programmes. A majority of the chapters come from Asian contexts, including Hong Kong (He & Lin, this issue; Lo, Lui, & Wong, this volume; Xu & Harfitt, this volume), Mainland China (An, Macaro, & Childs, this volume) and Singapore (Ho, Wong, & Rappa, this volume). In these contexts, English is the target language, but its status in the wider political and socio-economic contexts is different. In Hong Kong and Singapore, two former British colonies, English is one of the official languages. However, English is the sole medium of instruction in Singaporean schools and serves as the lingua franca among Singaporeans of different ethnic groups (Tan, 2014). On the other hand, Chinese (spoken Cantonese and Standard Written Chinese) is the language of daily communication in Hong Kong, where over 90% of the population is Chinese (Li, 2017). Therefore, English is the first or most familiar language (L1) of quite a significant proportion of students in Singaporean schools, but not in Hong Kong. On the other hand, English is largely learned and used as a foreign language in Mainland China. Despite a large number of English language learners, English is seldom used as a language for daily communication among Chinese people (Fang, 2017). However, in the era of globalisation, international trade and educational migration, Chinese people, especially the younger generations, have attached more importance to English (Bolton & Graddol, 2012). More and more international programmes (set up in public high schools) and international private schools have been established to provide EMI education for students who plan to study in overseas EMI tertiary institutes. These schools and programmes often follow a foreign curriculum taught by monolingual English-speaking teachers. Therefore, in the chapters contextualised in these three educational settings, the programmes that the researchers examine could be placed along the "content-driven" end of the continuum discussed above. In these studies, the researchers are proposing ways to move the programmes towards the ideal practices of CLIL along the continuum.

The remaining chapter (Turner, this volume) comes from Australia, where CLIL programmes targeting at languages other than English (e.g., Italian, Japanese, Mandarin) have been in place. Judging from Turner's description in the chapter, the programme she examines is more inclined towards the language-driven end of the continuum. Turner's chapter in turn constitutes an interesting empirical study in this volume and the field of CLIL research in general, in terms of the nature of the programme and the target language. More details of these six chapters will be discussed in the last part of this introduction.

2. Content and language integration in science

It is widely agreed that mastery of the content of a discipline is in large part mastery of the discipline's specific ways of using language, or discipline-specific literacy (Lemke, 1990; Schleppegrell, 2004). On one hand, this proposition justifies the CLIL approach, in the sense that content subject provides authentic and meaningful contexts for learning and using L2 (Genesee & Lindholm-Leary, 2013). On the other hand, it also explains why CLIL students often encounter difficulties, as academic and discipline-specific genres and registers differ from the everyday genres and registers that they are more familiar with.

Schleppegrell (2009) delineates the scope of academic language "as a set of linguistic *registers* that construe multiple and complex meanings at all levels and in all subjects of schooling," (p. 1; original emphasis) whereas Zwiers (2008) explicates it as "the set of words, grammar, and organisational strategies used to describe complex ideas, higher order thinking processes and abstract concepts" (p. 20). Unlike what most content subject teachers tend to believe, academic language is not restricted to subject-specific vocabulary, but also grammar features, sentence patterns and genres (Cammarata & Haley, 2018). The features of academic language have been analysed from different perspectives. Cummins (2000) contrasts basic interpersonal communication skills (or conversational language) and cognitive academic language proficiency (or academic language), seeing them as differing in terms of the cognitive demand and contextual support available. Academic language is often more cognitively demanding, without much contextual support. At the same time, researchers inspired by systemic functional linguistics identify the characteristics of academic language with reference to such concepts as communicative purpose (genre), mode, tenor and field (register). The language of science has been widely researched and it has been observed that some common genres in science include procedure, procedural recount (i.e., lab report), information report and explanation, which are characterised by genre-specific schematic structuring patterns and the prevalence of subject-specific technical terms, high lexical density, complex noun groups, and the use of passive voice (de Oliveira, 2010; Schleppegrell, 2004). Such a way of using language in science makes it a different register of language that even the so-called native speakers need to acquire separately, not to mention students learning in their L2 in CLIL programmes. Indeed, if one expands the scope of the "additional language" in the definition of CLIL to any language system or register that the students are not familiar with, then "academic language" in a sense can also be regarded as an additional language, since it is another register of language that everyone, regardless of whether they are L1 or L2 speakers, needs to master. In this sense, most chapters in this volume are actually addressing the target language at two levels, one being another language system or variety (e.g., English, Japanese), and the other being another register (e.g., academic registers).

Therefore, to achieve the dual goal of content and language integrated learning, students in CLIL science classrooms are expected to master several things simultaneously, including the abstract scientific knowledge and concepts, cognitive organising and argumentative skills, conversational language involved in classroom interaction, and academic language required to understand and express their science knowledge. It is then not surprising that students encounter difficulties in CLIL and one cannot assume that students can "pick up" all these learning targets. This is especially true in educational contexts where students may not have been sufficiently prepared for CLIL in terms of their L2 proficiency (e.g., in some English-as-a-foreign-language [EFL] contexts, where CLIL has become popular for its claimed benefits in L2 learning [Lo & Lin, 2015]). It is against such backdrop that more and more CLIL researchers and educators have been calling for more "integrated" teaching of content and language, or more explicit instruction of language (Cammarata & Haley, 2018; Llinares, Morton, & Whittaker, 2012).

However, integrating content and language teaching is easier said than done. The challenges can be understood at several levels. Some are closely associated with teachers' professional training and hence their beliefs and identities. It is observed that in most contexts, CLIL is mainly practised by content subject specialists (Dalton-Puffer et al., 2014; Wolff, 2012), although in some contexts, CLIL is practised by language specialists or by teachers with qualifications in both content and language subjects. With their professional training as content subject (e.g., science) specialists, some teachers may not fully understand the rationale behind CLIL or its potential benefits (Kong, Hoare, & Chi, 2011). They also tend to construct their identity as "content subject teachers" *only*, and they may not believe in their *dual* role of teaching both content and discipline-specific language (Lo, 2014). This may be particularly true for science teachers, who tend to privilege the notion of "concepts" and view mastery of science chiefly as mastery of science concepts without realising the role of lexicogrammatical resources in mediating these concepts (Seah, Clarke, & Hart, 2015).

Other challenges are related to the different pedagogies between content subjects and language teaching. As Dalton-Puffer (2013) highlights, "the problem of integration in CLIL needs be solved on the level of the different pedagogies … What is at issue, then, is the need to link up the pedagogies of the different subjects like mathematics, history or economics with the pedagogy of language teaching" (p. 219). Science pedagogy emphasises inquiry-based instruction, in which students are actively engaged in the process of exploring, asking questions, and constructing knowledge. Inquiry-based pedagogies may not be totally compatible with some strands of L2 pedagogies which encourage the pre-teaching of difficult vocabulary/terms (e.g., the SIOP model; see Echevarria, Vogt, & Short, 2013). Thus, one key challenge facing science teachers in CLIL is how they can systematically integrate content and language teaching in their lessons without frontloading too much language teaching that might interfere with the inductive, scientific inquiry process (Weinburgh, Silva, Smith, Groulx, & Nettles, 2014). The existing CLIL literature, however, tends to draw on mainly L2 pedagogies without adequately addressing the (potential) conflict between subject-specific pedagogies (e.g., science pedagogy) and L2 teaching pedagogies (Morton, 2016).

The past few years has witnessed a rapid growth in research on instructional practices and professional development in CLIL settings, as evident by several special issues on the topic in international journals (e.g., Cammarata & Ó Ceallaigh, 2018; Lyster & Ruiz de Zarobe, 2018). These publications have proposed new ideas and insights into CLIL curriculum planning and pedagogical practices (e.g., Cammarata, 2016; Lyster, 2016). This volume seeks to further contribute to the field by focusing on the notion of "scaffolding" – how science teachers can provide appropriate and timely scaffolding for their students to overcome the aforementioned challenges in CLIL science classrooms.

3. Scaffolding

The idea of "scaffolding" is closely related to the social constructivist theory of learning, which is developed based on the ideas proposed by Vygotsky (1978). In this perspective, cognitive and affective development, and hence learning, takes place in social interaction. When language is used in social activities (e.g., interacting and collaborating with others), it can serve as a mediating tool in all forms of higher-order mental processing such as logical problem-solving, planning and evaluating, learning, etc. (Swain & Lapkin, 2000). During interaction or collaboration, people are engaged in dialogue, where they can externalise their thoughts in the form of utterances. These thoughts can then be "scrutinized, questioned, reflected upon, disagreed with, changed, or disregarded" (Swain & Lapkin, 2000, p. 286). Through such social, interactional activity, which is called "intermental activity", learners can reconstruct their own understandings and ways of thinking, which in turn promotes learning within learners (i.e., "intramental activity") (Rojas-Drummond & Mercer, 2003). During those interaction processes, teachers or the more competent peers may provide assistance for learners in solving the problems or performing the tasks that may be beyond their ability. The difference between the learners' current level of development without any assistance and the potential level that learners could achieve with others' assistance is called the "Zone of Proximal Development" (Vygotsky, 1978), where learning can take place. Meanwhile, the process that enables a learner to accomplish a task or achieve a goal which would otherwise be beyond his efforts is called "scaffolding" (Wood, Bruner, & Ross, 1976). During this process, parents, teachers or the more competent peers normally do the following to help the learners (Cameron, 2001; Silver, 2011; Wood et al., 1976):

- understand the learners' existing level of knowledge and relate the new knowledge to what they already know or what they can do
- maintain the learners' interest in the task
- simplify the task, or break the task into smaller, more manageable steps
- model the skills required or demonstrate how to do the task
- provide hints, prompts or cues

From this social constructivist view of learning, scaffolding is learner-centred, catering for learners' existing level and needs so as to facilitate learning in their Zone of Proximal Development (Ankrum, Genest, & Belcastro, 2014). It is further argued that teachers should maintain high expectations of students, as long as they provide adequate scaffolding that is responsive to students' needs (Gibbons, 2015). Such "high challenge, high support" combination has significant implications for CLIL science classrooms, where, as previously described, students are likely to face high demands in content, cognitive and linguistic aspects. To help students to achieve the dual goal of content and language learning, teachers' scaffolding is crucial.

As the importance of scaffolding is well documented, the remaining questions to ask would be what, when and how. What kinds of scaffolds should CLIL teachers provide, especially considering the integration of content and language learning? And when and how should they provide these scaffolds in lessons? Some insights could be drawn from research on the English Language Learners (ELLs) in mainstream education in Anglophone countries (e.g., Echevarria, et al., 2013; Gibbons, 2015), which provides some concrete teaching ideas and lesson planning frameworks for incorporating scaffolding for ELLs. Gibbons (2015), in particular, demonstrates in detail how teachers can provide scaffolding when interacting with their ELL students, which echoes the importance of language and social interaction in the social constructivist view of learning.

Pawan (2008), after analysing over 3700 online postings by in-service content subject teachers in the US, proposes a model of different types of scaffolding (p.1454–55):

1. Linguistic scaffolding: Simplifying and making the English (target) language more accessible, for example, by shortening selections, speaking in the present tense, avoiding the use of idioms, direct instructions of writing and vocabulary, etc.
2. Conceptual scaffolding: Providing students with supportive frameworks for meaning by providing organisational charts, metaphors, visuals, modelling, experiments, etc. In other words, this type of scaffolding involves multimodality.
3. Social scaffolding: Using social interaction to support and mediate learning (e.g., group work, peer coaching, teacher providing one-to-one assistance)
4. Cultural scaffolding: Using artifacts, tools and informational sources that are culturally and historically situated within a domain familiar to learners (e.g., students' prior knowledge, literature from students' culture, L1 peer work)

Among these types of scaffolding, conceptual scaffolding was mentioned the most frequently by the participating teachers in Pawan's (2008) study (with nearly 50% of relevant postings), followed by social and linguistic scaffolding (with over 20% of relevant postings respectively). Yet, the teachers showed limited understanding and adoption of cultural scaffolding (only 6% of relevant postings). Such findings may not look surprising, as the participants were content subject teachers, who would inevitably put more attention to conceptual scaffolding. It is evident that they also paid some attention to linguistic scaffolding, which helps ELLs to overcome the language barriers, but they were urged to pay more attention to the cultural knowledge and experience that ELLs could potentially bring to the classrooms to facilitate their learning.

When it comes to CLIL, some studies, similar to Gibbon's approach (2015), focus on the verbal interaction between teachers and students so as to examine how teachers could support student learning. For instance, Lin and Wu (2015) present a fine-grained analysis of how a science teacher interacted with students, making use of L1, visual aids and gestures, to enable them to understand and describe the results of an experiment. Indeed, in the field of CLIL, more attention has been paid to the affordance of multiple resources of students' communicative repertoires, especially students' L1 resources (Lin & Lo, 2017; Nikula & Moore, 2019). Advocating the systematic planning and use of different resources and modes, Lin (2016) proposes a pedagogical framework, the Multimodalities-Entextualisation-Cycle (MEC), which incorporates different types of scaffolding (e.g., features of L1, L2, everyday registers, academic registers, visuals, experiments) at different stages of instruction to help students master content and L2 academic literacy. In a survey of over 200 CLIL teachers' self-reported pedagogical practices, van Kampen, Admiraal and Berry (2018) identified "scaffolding" as a key theme in the teachers' practices. 29% of the teachers surveyed believed they provided more scaffolding to students in their CLIL classes, compared with their regular classes. And such scaffolding included some linguistic ones (e.g., making glossaries for vocabulary learning) and conceptual ones (e.g., providing tools to analyse a text, using visuals to help with explanation).

Building on the foundation of these studies on scaffolding CLIL learners, this volume seeks to make further contribution on this important topic with a collection of six studies based in science classrooms.

4. Chapters in this volume

Chapters 2 to 4 in this volume focus primarily on teacher-student verbal interaction, during which teachers provide scaffolding for students by teaching form-meaning relationship explicitly. An et al.'s study (this volume) was contextualised in an emerging type of EMI secondary schools in Mainland China, which adopts the curriculum of Anglophone countries taught by monolingual, English-speaking teachers. The researchers focus on the notion of "Language Focused Episodes" (LFEs), during which science teachers temporarily drew students' attention to the language "forms" (e.g., vocabulary, grammatical structures, idioms), thereby providing linguistic scaffolding for students. In their chapter, An et al. demonstrate how monolingual science teachers interacted with students in the LFEs when explaining general academic vocabulary, morphemes and idioms. While these LFEs are interesting and potentially useful for integrating science and language learning, they were not extensively found in An et al.'s corpus of 30 science lessons, which led the researchers to discuss the implications for teacher education, especially in terms of raising the language awareness of monolingual teachers.

In Chapter 3, Turner (this volume) presents a rather unique CLIL science programme, in terms of the target language (Japanese) and the educational setting, which is an Anglophone country (Australia). The CLIL programme described in Turner's chapter appears to be rather language-driven, as the teachers had their own discretion on the proportion of using Japanese and English (L1), and students' science knowledge was assessed mainly in their L1. Another interesting point about the CLIL classroom in this chapter is that it was co-taught by two science teachers, one being monolingual English-speaker and the other having learned Japanese as a foreign language. With data from lesson observations and interviews with teachers and students, Turner illustrates how the language features of Japanese (*kanji*), the target language, could help reinforce students' understanding of some science concepts. This serves as a nice example of how language can facilitate content learning, especially in a language other than English. However, it was also observed that Japanese seemed to be marginalised as the language for warming-up and application stages of the science lessons, while English (L1) was regarded as the main language when exploring and explaining science concepts and in assessments. Turner then critically discusses the integration and separation of the two languages in that particular Japanese CLIL setting.

The notion of "teacher language awareness" (TLA) is implied in the first and second chapters, in the sense that teachers with a higher level of language awareness are likely to better integrate science and language teaching. Such a notion is then explicitly examined in Chapter 4 by Xu and Harfitt (this volume). Conducting their study in EMI secondary schools in Hong Kong, the researchers examine how science teachers translated their TLA into their pedagogical practices, especially in terms of "scaffolded interaction". The researchers have identified six scaffolding strategies in science lessons which are associated with different dimensions of TLA. Xu and Harfitt's chapter, therefore, contributes to our understanding of not only CLIL pedagogical practices, but also teachers' pedagogical content knowledge (Morton, 2016).

Chapters 5 and 6 extend beyond teacher-student verbal interaction and incorporate visuals as well. Ho et al.'s chapter (this volume) examines science learning in a secondary school in Singapore. With the analytical framework of "talk moves" (Chapin, O'Connor, & Anderson, 2013), the researchers demonstrate how a science teacher clarified students' misunderstanding of a very specific biological concept. The teacher-student interaction was also based on concept sketches, which serve as a visual representation in science. Ho et al. thus illuminate the potential of combining linguistic and other modes (e.g., visuals) to scaffold students' mastery of content and academic literacy.

In Chapter 6, He and Lin (this volume) draw on Lemke's (1990) notion of "thematic patterns" (which are the semantic relationships between key concepts in science) and develop what they call the "Concept + Language Mapping" (CLM) approach to integrate content and language teaching. Under this approach, some "Concept + Language" maps or cards are designed to help students understand key thematic patterns and master the associated academic language (e.g., sentence patterns). These maps and cards incorporate different semiotic resources, including language (L1, L2), pictures, graphic organisers, etc. These cards are then utilised by CLIL teachers, who are expected to engage in different interaction patterns with students so as to provide both "designed scaffolding" (i.e., the designed teaching materials) and "spontaneous scaffolding" (teacher-student interaction). This pedagogical approach was tried out in some biology lessons in a quasi-experimental design in an EMI secondary school in Hong Kong. Both quantitative and qualitative data reveal the effectiveness of this innovative pedagogical approach, thereby showing the positive impact of different types of scaffolding in CLIL science classrooms.

Chapter 7 (Lo et al., this volume) focuses primarily on linguistic scaffolding, which in a sense is similar to Chapters 2 to 4. However, the focus of this chapter is on scaffolding for assessment, an issue which has been rather widely discussed in studies on ELLs (Gottlieb, 2006; Mihai, 2010) but not yet in CLIL (Massler, Stotz, & Queisser, 2014). Intrigued by the complexity of assessing students' progress and difficulties in both content and L2 learning in CLIL, Lo et al. (this volume) set out to examine whether science teachers in EMI schools in Hong Kong aligned their objectives, instruction and assessment, considering the dual goal of CLIL. By comparing the cases of two science teachers, the researchers reveal how one teacher provided scaffolding in both cognitive and linguistic aspects through incorporating content and language objectives in the lessons, as well as highlighting some sentence patterns and text structures that were useful for students to express their science knowledge, especially in response to some typical questions in assessment practices. Their findings point to the importance of providing scaffolding, especially in the linguistic aspect, so that students' progress in content learning could be assessed in a more valid way.

Finally, the implications of the six empirical studies are brought together by Tang (this volume), who is a science educator. In this concluding chapter, Tang synthesises and discusses the findings of the papers from the lens and perspectives of a subject specialist, which is urgently needed but has been lacking in CLIL research (Dalton-Puffer, 2018).

We believe this volume, with studies from different educational settings and epistemological paradigms, and adopting a variety of research designs, will add to the growing knowledge in CLIL. Although this volume focuses on the science discipline, which undoubtedly has its unique features, we believe some key insights drawn from the studies, particularly those into scaffolding and content and language integration, can well be extended to other disciplines.

References

Ankrum, J. W., Genest, M. T., & Belcastro, E. G. (2014). The power of verbal scaffolding: "Showing" beginning readers how to use reading strategies. *Early Childhood Education Journal*, 42(1), 39–47. https://doi.org/10.1007/s10643-013-0586-5

Bolton, K., & Graddol, D. (2012). English in China today. *English Today*, 28(3), 4–9. https://doi.org/10.1017/S0266078412000223

Cameron, L. (2001). *Teaching languages to young learners*. Cambridge: Cambridge University Press. https://doi.org/10.1017/CBO9780511733109

Cammarata, L. (2016). Foreign language education and the development of inquiry-driven language programs: Key challenges and curricular planning strategies. In L. Cammarata (Ed.), *Content-based foreign language teaching: Curriculum and pedagogy for developing advanced thinking and literacy skills* (pp. 123–143). New York, NY: Routledge. https://doi.org/10.4324/9780203850497

Cammarata, L., & Haley, C. (2018). Integrated content, language, and literacy instruction in a Canadian French immersion context: A professional development journey. *International Journal of Bilingual Education and Bilingualism*, 21(3), 332–348. https://doi.org/10.1080/13670050.2017.1386617

Cammarata, L., & Ó Ceallaigh, T. J. (Eds.) (2018). Teacher education and professional development for immersion and content-based instruction: Research on programs, practices, and teacher educators [Special issue]. *Journal of Immersion and Content-Based Language Education*, 6(2). https://doi.org/10.1075/jicb.00004.cam

Cenoz, J., Genesee, F., & Gorter, D. (2014). Critical analysis of CLIL: Taking stock and looking forward. *Applied Linguistics*, 35(3), 243–262. https://doi.org/10.1093/applin/amt011

Chapin, S., O'Connor, C., & Anderson, N. (2013). *Classroom discussions in Math: A teacher's guide for using talk moves to support the common core and more, Grades K-6: A Multimedia Professional Learning Resource (3rd ed.)*. Sausalito, CA: Math Solutions Publications.

Coyle, D., Hood, P., & Marsh, D. (2010). *CLIL: Content and language integrated learning*. Cambridge: Cambridge University Press.

Cummins, J. (2000). *Language, power, and pedagogy: Bilingual children in the crossfire*. Clevedon, UK: Multilingual Matters. https://doi.org/10.21832/9781853596773

Dalton-Puffer, C. (2007). *Discourse in Content and Language Integrated Learning (CLIL) classrooms*. Amsterdam: John Benjamins. https://doi.org/10.1075/lllt.20

Dalton-Puffer, C. (2013). A construct of cognitive discourse functions for conceptualising content-language integration in CLIL and multilingual education. *European Journal of Applied Linguistics*, 1(2), 216–253. https://doi.org/10.1515/eujal-2013-0011

Dalton-Puffer, C. (2018). Postscriptum: Research pathways in CLIL/Immersion instructional practices and teacher development. *International Journal of Bilingual Education and Bilingualism*, 21(3), 384–387. https://doi.org/10.1080/13670050.2017.1384448

Dalton-Puffer, C., Llinares, A., Lorenzo, F., & Nikula, T. (2014). "You can Stand Under my Umbrella": Immersion, CLIL and Bilingual Education. A Response to Cenoz, Genesee & Gorter (2013). *Applied Linguistics*, 35(2), 213–218. https://doi.org/10.1093/applin/amu010

de Oliveira, L. C. (2010). Enhancing content instruction for ELLs: Learning about language in science. In D. Sunal, C. Sunal, M. Mantero, & E. Wright (Eds.), *Teaching science with Hispanic ELLs in K-16 classrooms* (pp. 135–150). Charlotte, NC: Information Age.

Echevarria, J., Vogt, M. E., & Short, D. J. (2013). *Making content comprehensible for English language learners: The SIOP Model (4th ed.)*. New York: Pearson.

Fang, F. (2017). World Englishes or English as a Lingua Franca: Where does English in China stand? *English Today*, 33(1), 19–24. https://doi.org/10.1017/S0266078415000668

Genesee, F., & Lindholm-Leary, K. (2013). Two case studies of content-based language education. *Journal of Immersion and Content-Based Language Education*, 1(1), 3–33. https://doi.org/10.1075/jicb.1.1.02gen

Gibbons, P. (2015). *Scaffolding language, scaffolding learning: Teaching second language learners in the mainstream classroom (2nd ed.)*. Portsmouth, NH: Heinemann.

Gottlieb, M. (2006). *Assessing English language learners: Bridges for language proficiency to academic achievement*. Thousand Oaks: Corwin.

Kong, S., Hoare, P., & Chi, Y.P. (2011). Immersion education in China: Teachers' perspectives. *Frontiers of Education in China*, 6(1), 68–91. https://doi.org/10.1007/s11516-011-0122-6

Lasagabaster, D., & Sierra, J.M. (2010). Immersion and CLIL in English: More differences than similarities. *ELT Journal*, 64(4), 367–75. https://doi.org/10.1093/elt/ccp082

Lemke, J.L. (1990). *Talking science: Language, learning and values*. Westport, CT: Ablex.

Li, D.C.S. (2017). *Multilingual Hong Kong: Communities, Languages, Identities*. Berlin: Springer. https://doi.org/10.1007/978-3-319-44195-5

Lin, A.M.Y. (2016). *Language across the curriculum and CLIL in English-as-an-additional-language contexts: Theory and practice*. Dordrecht: Springer. https://doi.org/10.1007/978-981-10-1802-2

Lin, A.M.Y., & Lo, Y.Y. (2017). Trans/languaging and the triadic dialogue in Content and Language Integrated Learning (CLIL) Classrooms. *Language and Education*, 31(1), 26–45. https://doi.org/10.1080/09500782.2016.1230125

Lin, A.M.Y., & Wu, Y. (2015). 'May I speak Cantonese?' – Co-constructing a scientific proof in an EFL junior secondary science classroom. *International Journal of Bilingual Education and Bilingualism*, 18(3), 289–305. https://doi.org/10.1080/13670050.2014.988113

Llinares, A., & Morton, T. (Eds.). (2017). *Applied linguistics perspectives on CLIL*. Amsterdam: John Benjamins. https://doi.org/10.1075/lllt.47

Llinares, A., Morton, T., & Whittaker, R. (2012). *The roles of language in CLIL*. Cambridge: Cambridge University Press.

Lo, Y.Y. (2014). Collaboration between L2 and content subject teachers in CBI: Contrasting beliefs and attitudes. *RELC Journal*, 45(2), 181–196. https://doi.org/10.1177/0033688214535054

Lo, Y.Y., & Lin, A.M.Y. (2015). Special Issue: Designing Multilingual and Multimodal CLIL Frameworks for EFL students. *International Journal of Bilingual Education and Bilingualism*, 18(3), 261–269. https://doi.org/10.1080/13670050.2014.988111

Lyster, R. (2016). *Vers une approche intégrée en immersion*. Montréal: Les Éditions CEC.

Lyster, R., & Ruiz de Zarobe, Y. (Eds.). (2018). Instructional practices and teacher development in CLIL and immersion school settings [Special issue]. *International Journal of Bilingual Education and Bilingualism*, 20(3). https://doi.org/10.1080/13670050.2017.1383353

Marsh, D. (2002). *CLIL/EMILE – the European dimension: actions, trends and foresight potential*. Finland: University of Jyväskylä. Retrieved from <https://jyx.jyu.fi/bitstream/handle/123456789/47616/david_marsh-report.pdf?sequence=1&isAllowed=y>

Massler, U., Stotz, D., & Queisser, C. (2014). Assessment instruments for primary CLIL: The conceptualisation and evaluation of test tasks. *The Language Learning Journal*, 42(2), 137–150. https://doi.org/10.1080/09571736.2014.891371

Mihai, F.M. (2010). *Assessing English language learners in the content areas: A research-into-practice guide for educators*. Ann Arbor, MI: University of Michigan Press. https://doi.org/10.3998/mpub.2825611

Morton, T. (2016). Conceptualizing and investigating teachers' knowledge for integrating content and language in content-based instruction. *Journal of Immersion and Content-Based Language Education*, 4(2), 144–167. https://doi.org/10.1075/jicb.4.2.01mor

Nikula, T., Dalton-Puffer, C., Llinares, A., & Lorenzo, F. (2016). More than content and language: The complexity of integration in CLIL and multilingual education. In T. Nikula, E. Dafouz, P. Moore, & U. Smit (eds.), *Conceptualising integration in CLIL and multilingual education* (pp. 1–25). Bristol: Multilingual Matters. https://doi.org/10.21832/9781783096145-004

Nikula, T., & Moore, P. (2019). Exploring translanguaging in CLIL. *International Journal of Bilingual Education and Bilingualism*, 22(2), 237–249. https://doi.org/10.1080/13670050.2016.1254151

Pawan, F. (2008). Content-area teachers and scaffolded instruction for English language learners. *Teaching and Teacher Education*, 24(6), 1450–1462. https://doi.org/10.1016/j.tate.2008.02.003

Rojas-Drummond, S., & Mercer, N. (2003). Scaffolding the development of effective collaboration and learning. *International Journal of Educational Research*, 39(1), 99–111. https://doi.org/10.1016/S0883-0355(03)00075-2

Schleppegrell, M. J. (2004). *The language of schooling: A functional linguistics perspective.* New York, NY: Routledge. https://doi.org/10.4324/9781410610317

Schleppegrell, M. J. (2009). Language in academic subject areas and classroom instruction: What is academic language and how can we teach it? Paper presented at workshop on *The role of language in school learning* sponsored by the National Academy of Sciences, Menlo, CA. Retrieved from <https://www.rcoe.us/educational-services/files/2012/08/What_is_Academic_Language_Schleppegrell.pdf>

Seah, L. H., Clarke, D. J., & Hart, C. (2015). Understanding middle school students' difficulties in explaining density from a language perspective. *International Journal of Science Education*, 37(14), 2386–2409. https://doi.org/10.1080/09500693.2015.1080879

Silver, D. (2011). Using the 'Zone' help reach every learner. *Kappa Delta Pi Record*, 47(sup1), 28–31. https://doi.org/10.1080/00228958.2011.10516721

Somers, T., & Surmont, J. (2012). CLIL and immersion: How clear-cut are they? *ELT Journal*, 66(1), 113–116. https://doi.org/10.1093/elt/ccr079

Swain, M., & Lapkin, S. (2000). Task-based second language learning: the uses of the first language. *Language Teaching Research*, 4(3), 251–274. https://doi.org/10.1177/136216880000400304

Tan, M. (2011). Mathematics and science teachers' beliefs and practices regarding the teaching of language in content learning. *Language Teaching Research*, 15(3), 325–342. https://doi.org/10.1177/1362168811401153

Tan, Y. (2014). English as a 'mother tongue' in Singapore. *World Englishes*, 33(3), 319–339. https://doi.org/10.1111/weng.12093

Van Kampen, E., Admiraal, W., & Berry, A. (2018). Content and language integrated learning in the Netherlands: teachers' self-reported pedagogical practices. *International Journal of Bilingual Education and Bilingualism*, 21(2), 222–236. https://doi.org/10.1080/13670050.2016.1154004

Vygotsky, L. S. (1978). *Mind in society: The development of higher psychological processes.* Cambridge, MA: Harvard University Press.

Weinburgh, M., Silva, C., Smith, K. H., Groulx, J., & Nettles, J. (2014). The intersection of inquiry-based science and language: Preparing teachers for ELL classrooms. *Journal of Science Teacher Education*, 25(5), 519–541. https://doi.org/10.1007/s10972-014-9389-9

Wolff, D. (2012). The European framework for CLIL teacher education. *Synergies*, 8, 105–116.

Wood, D. J., Bruner, J. S., & Ross, G. (1976). The role of tutoring in problem solving. *Journal of Child Psychiatry and Psychology*, 17(2), 89–100. https://doi.org/10.1111/j.1469-7610.1976.tb00381.x

Zwiers, J. (2008). *Building academic language: Essential practices for content classrooms, Grades 5–12*. San Francisco, CA: John Wiley & Sons.

Language focused episodes by monolingual teachers in English Medium Instruction science lessons

Jiangshan An[1], Ernesto Macaro[2] and Ann Childs[2]
[1] Purdue University Fort Wayne | [2] University of Oxford

This study is situated in a newly emerging EMI setting in China where an Anglophone high school curriculum is taught by predominantly foreign teachers through English to local Chinese students. These teachers are termed 'monolingual teachers' in the sense that they cannot use the students' L1 as a resource in their teaching should they wish to, as opposed to the typical bilingual teachers commonly explored in the existing EMI literature. Through quantitative and qualitative analysis of 30 video-recorded EMI science lessons taught by 15 monolingual teachers we identified and explored the language-focused-episodes (LFEs) where students' attention was explicitly diverted from the content plane to the language plane. We found very limited explicit language instruction, with non-technical vocabulary being the main type of LFEs, and only a narrow range of grammatical features being attended to. The implications for this lack of focus on language are discussed in the context of monolingual teachers but also with reference to the potential for bilingual teachers to use both L1 and L2 for LFEs.

Keywords: bilingual education, classroom interaction, content-based instruction, discourse analysis

1. Introduction

In recent decades the world of education has seen an unprecedented growth in classrooms where academic subjects (or 'content subjects'), such as science, mathematics, geography and business studies, are taught through a language which is not the first language of the students in that classroom, or at least not of the majority of the students (Coleman, 2006; De Graaff, Koopman, Anikina, & Westhoff, 2007; Lorenzo, 2007). In North America these classrooms are often designated 'immersion' classrooms (Cammarata & Tedick, 2012). In Europe, they are almost invariably designated Content and Language Integrated Learning (CLIL) classrooms (Pérez-Cañado, 2012). In Asia (but also in other parts of the world) they are very often designated English Medium Instruction (EMI) classrooms. EMI is a feature of both secondary and tertiary education (Macaro, 2018). Although these different classroom settings have different characteristics, one characteristic which almost invariably unites them is that they are taught by bilingual teachers (see below for a definition). However, this study fills a gap in the literature by looking lessons being taught by monolingual teachers who cannot use their students' L1 in their teaching.

2. Literature review

2.1 Growth of EMI

The worldwide growth of classrooms in the secondary phase of education where content is taught through the medium of predominantly English can be attributed to many factors among which: a belief that teaching content through an L2 can be a motivating factor (Marsh, 2002); that the presumed high levels of exposure to the L2 are beneficial to language learning (Nikula, Dalton-Puffer, & García, 2013) or to plurilingualism (Lasagabaster & Huguet, 2007); an expectation that when the L2 is learnt through the medium of content then it fulfils an often stated desire for language teaching to be carried out through a communicative methodology (García Mayo & Basterrechea, 2017); that these types of classrooms achieve a dual purpose of teaching both content and language and therefore are cost-effective. As the focus of the current study is not on tertiary education, we mention only briefly the EMI motivators of increasing a university's prestige and attracting lucrative international students. However, we should not lose sight of the possible interrelationship between the two phases of education (Macaro, Curle, Pun, An, & Dearden, 2018; Shohamy, 2006) and the relationship between private and state sector secondary education where teaching through L2 English is seen as a sign of prestige and potential advancements by parents. As our educational setting is one designated as EMI we will use this term henceforth as an umbrella term unless trying to make a comparison with other terms.

2.2 Teachers in EMI settings

As mentioned in the introduction, EMI classrooms are characterized by bilingual teachers. In other words, these are teachers who speak the first language (L1) of the students, either as native speakers or as highly proficient speakers of that language, as well as being highly proficient in L2 English.[1] By contrast, monolingual teachers are those that do not speak the L1 of the students to any degree of pedagogical functionality. We prefer the term 'monolingual teachers' to 'native speaker teachers' because of the connotations attached to the latter (see Llurda, 2005) but also because there may be non-native speakers of English, although they may be highly proficient, who might find themselves in a teaching situation in which they are monolingual teachers. Of course, monolingual teachers may well be bilinguals or even multilinguals. We define them as monolingual teachers simply on the basis that they do not speak the L1 of the students, such as the teachers in our study, and because of this are not able to use that L1 as a resource in their teaching should they wish to.

The advantages and disadvantages of not being able to speak the L1 of their students have been well documented in the literature (Cook, 1999; Llurda, 2005) and it is not our intention to enter into that discussion at this stage (but see Discussion). However, an obvious resource that bilingual teachers have is that they can make a link to the L1 for their students when a communication problem arises rather than providing a paraphrase or circumlocution and so on. A possible disadvantage (Butzkamm, 1998; McMillan & Turnbull, 2009) is that the L1 becomes the easy way out for solving a communication problem rather than through the 'modification of L2 input' (Krashen, 1982) or 'negotiation of meaning' in the L2 (Long, 1996).

2.3 Distinctive features of Chinese bilingual schools

A recent phenomenon in China has been the rise of foreign high school programs where a curriculum of an anglophone country is followed and often taught by a high number of foreign monolingual teachers. These programs often exist in what are termed 'international departments' located within public high schools, or in independent private schools. Students attending the international departments typically plan to transition to tertiary education in an Anglophone country.

---

1. One anonymous reviewer pointed out that in Content Based Instruction settings teachers often are monolingual as in our definition. A reason for not including CBI in the discussion or not comparing the literature in CBI and EMI is that the CBI and EMI contexts are very different in terms of: the quantity and quality of L2 exposure, as CBI students have contact with L2 (the majority language) in their everyday life where EMI students almost invariably do not; EMI students often do not have all subjects taught through English; the student body in EMI classrooms also often share the same L1 whereas CBI classes usually have students from different linguistic backgrounds (see Met, 1999; Macaro, 2018; for a discussion of the distinction between EMI and CBI).

In these foreign high school programs, academic subjects are partly or wholly taught through the medium of English. Examples of the curricula taught include the Canadian provincial high school curriculum (e.g., British Columbia, Alberta), IGCSE, A-level, IB, AP. There are often Chinese nationals teaching on these programs, but they also attract foreign monolingual teachers, such as teachers from Canada, the UK and the US, possibly as a consequence of the curricula and the students' motivation for studying abroad. In a number of countries (e.g., South Korea, Qatar, see Macaro, 2018) prestige is attached by schools, often in the private sector, to the fact that the curriculum is adopted from a 'Western' education system and that both the L1 and the local culture are not referred to in any significant way. The monolingual teacher population in the foreign high school programs (explored in this study), who cannot support students' L1 development, constitutes a distinctive feature of the EMI scenario, compared to the bilingual teachers commonly explored in the existing EMI literature.

The teaching and the curriculum that forms the context of our study in these schools is science (i.e., Biology, Physics, Chemistry) and it is to the implications of teaching science through EMI in the secondary phase that we now turn.

2.4   The language of science and science explanations

Teacher talk in science is predominantly focused on explanations. These can involve explanations (Miller, 2009) of a substance or an element (e.g., bauxite), of processes (e.g., to displace), and of apparatus (e.g., beaker). Some of these explanations then involve terms which are used across scientific disciplines (e.g., synthetic), everyday but polysemic words (e.g., to model), compounds and collocations (e.g., states of matter) and so on. The language of science subjects tends to typically express abstract concepts, logical relationships and universal phenomena (Lemke, 1990; Seah, Clarke, & Hart, 2014). It is also specialized through technical terms and specific patterns of discourse, such as lexical density, grammatical metaphor or nominalization, and syntactic discontinuity (Fang, 2005; Halliday, 1993; Martin, 1991). These linguistic devices, such as nominalization, may be particularly challenging because they involve a process of condensing into a technical term a great deal of underlying language, such as 'resultant force' in physics which is a highly condensed series of processes and actions.

Clearly then in EMI science classes, a student is faced not only with an understanding of the scientific concept through its explanation but also of the lexical items and phrases in L2 which are contained in that explanation. Therefore, when EMI studies report that students have problems with technical vocabulary (Yip, Tsang, & Cheung, 2003), it is of some empirical interest to identify further whether students mean problems with learning the form and surface meaning of the actual technical word, or they mean understanding the language of the explanation of the technical word. Whilst the boundaries between technical and non-technical lexical items are not watertight, we can define non-technical vocabulary as that which is unlikely to be listed in a technical dictionary or the glossaries of a textbook in a specific academic subject. General academic terms can be based on Coxhead (2000)'s academic words.

In the light of the above linguistic complexities it is of some importance to ask not only what kinds of explanations of scientific terms EMI teachers, and in this case monolingual EMI teachers, give but also the extent to which they might shift the focus of their explanations from the 'content' of science to the language itself. Of course, content and language are very often inextricably linked and hence the notion that CLIL will develop not only content but also language. However, part of our aim was to establish whether it was actually possible, from a corpus of EMI teachers' explanations, to differentiate when a teacher was purely explaining about a scientific concept from when s/he was focusing on an aspect of the English language, which we have termed 'Language Focused Episodes' (LFEs) in this study.

In terms of the difference between an LFE in an EMI class and a class in which the content is being taught through the students' L1, we could hypothesise that in L1 medium instruction classes the teacher would not have to explain to 14-year-old students the meaning of, for example, the word '*compile*' – it should be well established in their general academic vocabulary. However, for EMI students they may not indeed know the meaning of '*compile*' and thus an LFE might be something like '*compile means put together, and we often say compile a list of something*'.

2.5 Research on language focused episodes in SLA and in content contexts

In Second Language Acquisition research there has been a great deal of interest in focus on form (Williams & Doughty, 1998), which is the extent to which teachers focus through their interaction on the morphosyntax of the language. An emerging focus in SLA has been on focus on lexical form (Laufer, 2005) where the interest is on whether SLA teachers focus on the meanings, and to some extent the forms of words. Less research to our knowledge has been dedicated to focusing on lexical form or morphosyntax in the EMI context. Basturkmen and Shackleford (2015) found few language-related episodes in the interaction of university accounting classrooms and that these were mostly in relation to technical vocabulary. By analysing a corpus of lectures in EMI science subjects, Costa (2012) also found a very limited number of, what she refers to as, focus-on-form episodes. Given the lecture format, not many of these were responding to students' lack of comprehension but were possible language problems pre-empted by the teacher. Moreover, in some of these focus-on-form episodes the practice of codeswitching was present because these were bilingual (Italian/English) EMI teachers, unlike the monolingual teachers in our study. Thus, in this study, featuring smaller secondary school classrooms taught by monolingual teachers, the extent to which LFEs occur and the language aspects addressed are explored.

The issues raised in previous literature led us to formulate the following research questions:

1. To what extent do monolingual science teachers in an EMI secondary school setting provide language focused episodes (LFEs) for their students?
2. What kinds of LFEs do they provide?

3. Methodology

3.1 Study design

This study adopted a mixed methods approach (Gorard & Taylor, 2004). The total time length and time proportion of the total and different types of LFEs in 30 lessons collected from the sample schools (see below) were counted to provide quantitative results in order to address the first research question. The content of LFEs is also described and commented on in order to address research question 2.

3.2 The population

The target school population were students and their monolingual teachers in foreign high school programs in China where a foreign curriculum is taught by predominantly monolingual teachers and the students are a homogenous group of Chinese students, who typically plan to go overseas for tertiary education.

3.3 The sample (schools, teachers and students)

We found a lack of official documentation about the number, distribution and background of the foreign high school programs in China, making random sampling difficult. Convenience sampling was therefore used and consistent effort was made to ensure the schools selected could represent variations of a range of characteristics of the target school programs, such as the geographical location, school size, the type of curriculum being taught, the academic attainment and English proficiency of the students. This was to ensure a reasonable level of representativeness of the sample to the population.

The students' academic and English performance was also difficult to obtain as there were no standardized public exams which applied to the majority of such programs in China. Therefore, teachers were asked to comment on the level of students' academic attainment and English proficiency and they confirmed that there was a spread of students at different levels. Because of the high tuition fee of these foreign programs, students were most likely to come from affluent family backgrounds.

Seven schools in five cities from three provinces were recruited for this study, involving 15 foreign monolingual teachers. All of them specified in a background questionnaire that they did not have a functioning level of Mandarin and their most proficient language was English.

The sample of teachers had a range of past teaching experiences regarding the years of teaching, subjects taught and the context they taught in. The overall characteristic is that most of them had taught English as a foreign language outside of their home country, and taught science subjects in their home country, i.e., L1 English context (e.g., Canada), often with students of ESL background in their classes. Only a few had previous EMI teaching experience, e.g., teaching Chemistry in English to Thai students in Thailand. Most of them had a bachelor's degree in science and a science teaching qualification in a majority English speaking country, such as Canada, the US, the UK. All the teachers referred to in this study were represented by a number, such as T13, to maintain anonymity. An Appendix is attached to provide more background information about the teachers in the sample.

3.4 Data collection

Non-participant observation was carried out by the first author. Two lesson observations were conducted for each of the 15 teachers, making 30 lessons in total, and were video recorded with participants' consent. The lessons were typically between 45 minutes to one hour, making 1520 minutes of lessons observed and recorded, and topics taught ranged from plant structure, biome types, acidic and basic solution, to sound waves and so on. Among the 30 lessons observed, one was in a lab classroom where the students were conducting experiments in groups with limited teacher-whole class interaction time. All the other 29 lessons were in a traditional classroom setting with the teacher standing in the front and no lab equipment in the classroom.

3.5 Data analysis

All the 30 lessons were transcribed verbatim. Transcripts were imported to NVivo and all LFEs were identified and coded. The types of LFEs were also coded according to the linguistic aspect addressed in an inductive manner as the types of linguistic features emerged from the data. 10% of the lessons (i.e., 3 lessons) were randomly selected to be coded again by the first author more than three months after the lessons were first coded, following the same criteria of identifying LFEs and the different types of LFEs. This resulted in an intra-rater reliability of 0.82, indicating a reasonable level of reliability of the coding in this study (Robson, 2002). In addition, all three authors also spent a considerable amount of time examining a selection of the LFEs in order to ensure that they had been coded correctly.

The time length was used as the measurement of the quantity of LFEs rather than the number of occurrences. This is because each LFE varied in the effort and attention the teacher dedicated to it and the level of detail of the target linguistic feature being explored. Thus, the number of occurrences would not be an appropriate measurement to answer the first research question. It was decided the time length would be a reasonable way of quantifying and representing the effort that was put in to provide explicit language instruction. The time length of each LFE was produced in NVivo.

4. Findings

This section first presents the quantitative results to answer research question 1 regarding the extent to which LFEs were provided in the 30 EMI science lessons by monolingual teachers and part of research question 2 through showing the quantity of different types of LFEs. The second half goes on to present the qualitative results to address research question 2 with examples of the different types of LFEs.

4.1 The proportion of the total interaction and the different types of LFEs

Overall there were 47.1 minutes of LFEs in the 30 lessons observed and occupied 3.1% of the total teacher-whole class interaction time. This indicates that explicit teaching of the English language occupied a noticeable but relatively small proportion of the teacher-whole class interaction time.

Three domains of linguistic features addressed in the LFEs emerged – vocabulary, grammar and idioms. The vocabulary-focused LFEs include the teachers exploring the meaning of non-technical vocabulary, the non-science meaning of technical vocabulary (e.g., *conductor* on a train and *conductor* for electricity), pronunciation, and spelling. The grammar-focused LFEs include the teachers exploring morphemes. The idiom-focused LFEs concern the teachers explaining the meaning of idioms. The time length and percentage of each type of LFEs in the 30 lessons are presented in Table 1.

Table 1. Quantitative results for different types of LFEs

Types of LFEs		Length in minutes	Percentage in the overall LFEs
Vocabulary-focused LFEs	Meaning of non-technical vocabulary	31.4	66.7%
	The non-science meaning of technical vocabulary	3.2	6.9%
	Pronunciation	1.8	3.9%
	Spelling	1.1	2.4%
Grammar-focused LFEs	Morpheme	9.0	19.1%
Idiom-focused LFEs	Idiom	0.5	1.1%
Total		**47.1**	**100.0%**

Most of the LFEs concerned vocabulary, with the meaning of non-technical vocabulary occupying 66.7% of the total time of LFEs. 6.9% of LFEs introduced the everyday meaning of technical vocabulary by differentiating it from its scientific meaning. Morpheme was the second most prominent type, occupying 19.1% of the overall LFE time.

4.2 Examples of the different types of LFEs

We first give examples of the vocabulary-focused LFEs and then move on to examples from the grammar-focused episodes and finally end by looking at examples of idiom-focused LFEs.

4.2.1 *Vocabulary-focused LFEs*

a. *Meaning of non-technical vocabulary*

Explanation of non-technical vocabulary was frequently observed in this study, occupying as much as 66.7% of all LFEs. Examples include *spread, strip, potential, shimmering, neutral, hose, anchor, blend, separate, enhance, reverse, erupt, substance, transfer, indicate, conserve, equivalent, frigid, display, combine, universe* and so on. While some of these are everyday non-academic vocabulary, such as *hose, shimmering, anchor, strip, erupt*, some are in the Coxhead's (2000) academic word list, such as *enhance, neutral, reverse, transfer*. Very often the teachers initiated the explanation through comprehension checks (e.g., "do you know the meaning of XX?").

Extract 1 is from T13's first Biology lesson on sea floor spreading. In Extract 1, during the teacher's explanation of how sea floor spreading takes place, she paused the introduction of science content and used a comprehension check to see if students knew the word *eruption*, a case of teachers' explicitly diverting students' attention to the meaning of a non-technical word.

Extract 1.

Turn	Timespan	Content (target word: eruption)	Speaker
10	16:53–17:38	Ok? So, the main thing is that you can know this information right here *[Teacher pointing at a specific paragraph in the textbook]* because it is going to relate to sea floor spreading. Ok? So, sea floor spreading is proposed by um a scientist named Harry Hess from the United States of America. And basically, he realized that the earth is like a big bar magnet with magnetic waves surrounding it. So just like the picture that I've showed you earlier, kind of like that. Ok? Um he also knew that the ocean floor was being formed from eruptions under the sea. Now what's, what does the word eruption mean?	T
12	17:39–17:41	[wait time]	
13	17:42–17:48	Eruption means to? What does this word mean?	T
14	17:48–17:49	Burst.	S2
15	17:49–17:52	To Burst, yeah! To burst. To erupt. [Teacher using hand gestures to show "burst" and "erupt"]	T
16	17:52–18:14	Ok? Um because Harry has um knew that there's an eruption in the ocean floor. Uhm, this eruption is producing magma.	T

Here, the teacher was explicitly checking students' understanding of the word *erupt* and Student 2 was able to provide a synonym *burst*. We do not know of course if other students knew the meaning of the word 'burst' because no comprehension check of it was made to the whole class on this lexical item. To L1 students, *erupt* is probably a common word that they encounter outside of the science context and thus may not need an explicit semantic explanation. However, because of the EMI students' L2 language needs, such non-technical vocabulary was observed to be often explained.

Extract 2 below is another example, which comes from T7's first Biology lesson on the structure of plants. This is a rare case of student-initiated LFEs as opposed to the common case of teacher-initiated ones. Here, a student interrupted the teaching of the science content and raised a clarification request. In response, the teacher provided a lengthy explanation about the non-technical word *anchor*.

Extract 2.

Turn	Timespan	Content	Speaker
5	2:39–2:58	So for roots, it is a plant part, a part of a plant. This is the part that is typically, usually below the ground that often we cannot see. Some cases we can, though. [read aloud from the slides] *"They absorb water and dissolve minerals. It helps to anchor the parts that are above the ground, like the shoot system. And they often store food."* So the root system is a very important part of plants. [T paused for 10 seconds for students to comprehend].	T
6	2:58–2:59	Ms XX (the teacher's name)?	S1
7	2:59–3:00	Yes?	T
8	3:00–3:03	What is, emm, anchor? What's that?	S1
9	3:03–3:04	Anchor?	T
10	3:04–3:05	Yes.	S1
11	3:05–5:21	So, anchor means to, well, first, you can have a boat anchor. [T drew on the blackboard] So you drop it off the boat, it goes into the water, and it keeps your boat in one spot. An anchor is something that holds another object down. So to anchor a boat, you are holding it, you are keeping it in one spot by dropping an object that is very heavy into the water. It sinks to the bottom, it sits on the sand, and it holds your boat in that one spot. So this could be the sand. So when you drop, your boat's anchor is attached to a very long rope. A machine lets the anchor go into the water. It sits on the sand, and it's strong enough to keep your boat in this area, so that you can either stop and fish, you can get out of your boat to go around, and your boat won't float away. So for a root system, the roots help anchor the pieces above the ground to the soil, to another tree.	T

Turn	Timespan	Content	Speaker
		[T drew a flower on the blackboard and pointed to it as speaking] We have a flower. Here is the ground. You have your roots. These roots are going to help anchor this flower to the ground. So it's going to anchor the flower to help keep the flower in one spot. Without the roots, the flower is not going to be able to stay in the soil. It can fall. It can be blown away by the wind. So these roots keep it planted in the soil, in the ground.	
12	5:21–5:23	Does that make more sense?	T
13	5:23–5:25	Yeah, yes.	S1
14	5:25–5:30	So, when you think of anchor, think of something like a root that keeps a plat in one spot, or that holds it into the ground.	T

In this excerpt, a student initiated a question about the meaning of *anchor*, a non-technical word used in the teacher's definition of a biology concept, roots, in Turn 5. The teacher used a variety of input modification techniques that are often found in negotiation of meaning instances where communication breakdown occurs among interlocutors involving at least one nonnative speaker, e.g., exemplifying, drawing, paraphrasing. This is a process that has been extensively researched to be beneficial for L2 acquisition (Long, 1996; Pica, Young, & Doughty, 1987). Again, it is speculated that in L1 non-EMI classes, the word *anchor* would almost certainly be a known word to students at the high school level and teachers probably would not need to explain it, particularly not at this level of detail.

It is also interesting that while this lexical LFE took up two and a half minutes, a direct Chinese equivalent exists, which is "锚", with the same literal and figurative meaning as the English word has. A bilingual teacher would have been able to provide it. On the other hand, this example of negotiation for meaning would not have occurred if the direct equivalent had been immediately given.

b. *The non-science meaning of technical vocabulary*

Apart from the instruction on the meaning of non-technical vocabulary, we also found that, when a science term has an everyday non-science meaning that is different from the science meaning, the everyday sense was sometimes also explained to the students in detail. Extract 3 comes from T2's first Physics lesson where the teacher was introducing a list of new lexical items for the next chapter on Electrostatics. Here, the teacher provided a monologue to explain the meaning of *charged objects* where he explicitly explained the everyday (or alternative) meaning of *charge* with an example.

Extract 3.

Turn	Timespan	Content (target word: charge)	Speaker
20	5:38–6:23	Charged objects – let's talk about it – charged object. What does charge mean to you? That means when your father gives you his credit card, and you go to stores, go into the stores, shopping, and you're charging on the credit card. You know? You don't get to use a credit card? This is a different kind of charge, yes? Well uh we'll learn more about it, but a charged object is an object that has an electrical charge and electrical field. It has more electrons than protons or less electrons than protons. We'll get to the idea of it but it has an electrical charge, positive and negative.	T

In this LFE, the teacher explained the non-science meaning of *charge* through an everyday example of using a credit card, contextualising it through repetition and paraphrasing (e.g., *go to stores, go into stores, shopping*). Again, this demonstrates some input modification techniques which are often used in both instructional and naturalistic conversations to make input more comprehensible to non-native speaking interlocutors (Ellis & He, 1999). It is speculated that, although explicit instruction to distinguish between everyday meaning and science meaning of technical vocabulary is also important for English L1 students, explanations of everyday meaning may need to be more explicit for EMI students. This issue of clarifying the everyday meaning may also be more pressing in the EMI context where students may not be aware of the polysemous nature of certain technical vocabulary and thus not aware of the distinction.

c. *Pronunciation*

With 3.9% of the LFEs concerning pronunciation, Excerpt 4 provides an example taken from T3's second Biology lesson on the characteristics of different types of biomes. The context of this excerpt is that the class was asked to compare two given biomes. The student here was presenting his idea on the similarities between tundra and permanent ice.

Extract 4.

Turn	Timespan	Content (content word: permanent)	Speaker
10	1:45–1:56	First of all, the same of the tundra and perman…preman…	S5
11	1:56–1:57	Per-man-nent. Good.	T
12	1:57–2:29	Permanent ice. Em…first of all, their…em…location [wait for 8 seconds] They are both at the frigid zone. [S6 writing 'at the frigid zone' on the blackboard].	S5

In this example, the student was having difficulty in pronouncing the word *permanent*, an everyday non-technical word. The teacher provided the correct pronunciation with pauses between syllables to make it clear to the student, who then repeated the correct pronunciation. Again, this recast appears to be a decision by the teacher on the basis of the student being in an EMI context. Whether he was just trying to scaffold the communication, i.e., helping it along, or whether he was concerned with ensuring that mis-pronunciation would not happen again, we do not have sufficient evidence to provide a definitive answer.

d. *Spelling*

2.4% of the total LFEs in the 30 lessons focused explicitly on the spelling of vocabulary. Extract 5 comes from T2's first Physics lesson where he was going through a list of key vocabulary for the next chapter and focused on both the meaning and the spelling of *vary* and *very*.

Extract 5.

Turn	Timespan	Content (target word: variable)	Speaker
17	13:20–13:35	If somebody needs constant, what is it? If it's constant? Is this the same – can you give me an opposite of constant? A word? A single word that would be an opposite of constant? What? We've had independent and responding –	T
18	13:35–13:36	Variable.	S1
19	13:36–13:50	Variables. Very good. That comes from the word vary. Not V-e-r-y, meaning a lot of, but v-a-r-y, to change. Ok?	T

In turn 19, the teacher showed his awareness that the two non-technical words, *vary* and *very*, may be confusing to the L2 students due to either the distinction between the two vowels in the written form or the subtle distinction in the two phonemes. Apart from the focus on spelling, this LFE also contains the teacher's comprehension check on students' understanding of the meaning of *constant*.

4.2.2 *Grammar-focused LFEs*

In the grammar-focused LFEs, attention to morphemes is the only type that was found.

a. *Morphemes*

Morphemes were the second largest type of LFEs in this study, occupying 19.1% of the total LFEs. The morphemes that were addressed are mostly prefixes and suffixes in technical science vocabulary. Extract 6 comes from T4's first Biology lesson on the topic of chemicals of life (e.g., carbohydrates, lipid and protein). In this case, the teacher was focusing on the meaning of prefixes.

Extract 6.

Turn	Timespan	Content (target morphemes: poly, mono)	Speaker
210	18:30–18:38	So poly means what, Kathleen? "Poly" means what?	T
211	18:38–18:42	Er, something mixed together.	S1
212	18:42–18:47	Mixed together? "Poly" means many,	T
213	18:47–18:49	Mono means what?	T
214	18:49–18:50	One, one.	S1
215	18:50–18:57	One. Ok, so polymer is made of many monomers.	T

It could be argued that the morphemes also constitute the meaning of the science terms and thus these LFEs should be seen as content-focused. However, the meaning of prefixes and suffixes are not limited to the science terms but are linguistic knowledge that could be applied to other vocabulary outside of science learning, e.g., the use of *mono* in *monolingual*. It is an interesting question to ask whether the meaning of *poly* and *mono* would be common knowledge to L1 students at the high school level.

4.2.3 *Idiom-focused LFEs*

Explanations of idioms were also present in this study. In Extract 7 from T2's second Physics lesson, the students were given scissors and tape to conduct an experiment in pairs on how friction separates changes. They were just starting a new chapter on Electrostatics.

Extract 7.

Turn	Timespan	Content (target idiom: don't run with scissors)	Speaker
40	20:11–21:30	[Teacher handing out scissors and tapes to students.]	T
41	21:30–21:33	Don't run with scissors. Don't run with scissors. Can you figure out what that means?	T
42	21:33–21:35	Uh, you will hurt yourself.	S8
43	21:35–22:01	You can hurt yourself, yes. If you fall down on the scissors and they are sharp, right, you could hurt yourself. And that's the literal meaning. But the figurative meaning, I don't know if you use the words in Chinese, is, be careful. Just be careful. Doesn't mean you might not even have scissors, right? But you say I am	T

Turn	Timespan	Content (target idiom: don't run with scissors)	Speaker
		going to talk to my Chinese literature teacher and I am going to tell him what I think. I might say, be careful, don't run with scissors.	

In this excerpt, the teacher, who is from Canada, introduces the meaning of the idiom, *don't run with scissors* (which also means 'be careful' more generally). It is interesting that this is one of the very few instances in the corpus that the teacher actually wondered what the phrase might be in Chinese in order to make a local contact with the students. Again, we note that this LFE is not directly related to the science teaching.

It is also interesting to point out that the teacher used not only an inner circle idiom (Kachru, 1985), but also one that is typically used in Canada. This example also reflects that, given the varieties of English that exist in different regions of the world, the EMI students could be exposed to a particular variety of English, especially the non-academic use of English, depending on the background of the teacher.

5. Discussion

This study explored the explicit instruction of language in science classes in a newly emerging and under-researched EMI scenario in China featuring foreign monolingual teachers who do not share the students' L1. We termed individual occurrences of this explicit instruction Language Focused Episodes. The aim was to examine the extent to which explicit instruction of language was provided and the aspects of language focused on.

5.1 The extent of LFEs

The key findings show, first of all, that it was possible to differentiate between when a teacher was purely explaining a scientific concept and when s/he was explaining an aspect of the English language. A number of LFEs were identified where the teacher was explicitly directing students' attention from the content plane to the language plane. However, the small proportion of LFEs to the overall teacher-whole class interaction time suggests the extent to which the monolingual teachers shifted their explanation from content to language was limited.

This finding aligns with what was very often found in the bilingual EMI teachers' classes at different educational levels. For example, an absence of focus on form episodes was reported in the CLIL primary and secondary school classes in Pérez-Vidal's (2007) study in Catalonia, Spain. Similarly, only occasional explicit effort in teaching language was observed in EMI secondary school classes in Hong Kong (Lo, 2010), and a total of 76 occurrences of Focus on Form episodes in 16 hours of science lectures in Costas' (2012) study in Italy, where a sizable proportion of it was the explanation of technical vocabulary and thus could be considered focusing on content rather than language by the definition of LFEs in this study. The same limited explicit instruction on language was also reported in the body of work on teachers' oral feedback to students' erroneous output in EMI classes., e.g., the rare use of metalinguistic cues, explicit correction (e.g., Llinares & Lyster, 2014).

The need for explicit teaching of language within communicative language classrooms has been extensively discussed in SLA (e.g., Norris & Ortega, 2000; Spada & Tomita, 2010). In the context where content is instructed through an L2, such as EMI, a number of scholars have also called for the need for teachers to proactively plan their lessons in ways that create opportunities for students to notice and pay explicit attention to language forms during the learning of subject content (De Oliveira & Schleppegrell, 2015; Lyster, 2007; Pérez-Vidal, 2007; Swain, 1996), particularly after the Canadian French immersion students were found to lack accuracy in their receptive language skills. Thus, the limited amount of LFEs in this study may indicate that a more language-focused approach is needed for students' L2 development.

5.2 The variety of linguistic features focused on

A range of different types of LFEs were observed in the 30 EMI science lessons and revealed the monolingual teachers' frequent attention, primarily in a preemptive manner, to various aspects of non-technical vocabulary, including meaning, pronunciation, and spelling. Cases were also observed where the everyday meaning of technical vocabulary was explicitly pointed out with input modification techniques commonly used for second language learners (Ellis & He, 1999). This further demonstrates the non-technical everyday vocabulary being a major area of language focus and counters the notion that it is only technical vocabulary which causes students difficulties.

The considerable attention to non-technical vocabulary in a way reflects the difference between EMI and non-EMI settings. EMI settings are much more likely to require explanations of non-technical vocabulary, at least at the high school level, as this study indicates. It also reflects that in the EMI context, due to the limited exposure to L2 that students typically have outside of the school context, non-technical vocabulary which is usually common to L1 students could potentially cause problems for EMI students. The findings of this study reveal that these non-technical lexical items and phrases contained in the science explanations can inhibit students' comprehension of science content. As Extract 2 earlier shows, the surface literal meaning of the science concept *root* should not be challenging to L2 students at the high school level. However, a student asked for the meaning of the word *anchor*, which was contained in the science explanation and should be a known word to L1 students. This echoes with Rollnick's (2000) work on learning science through English as an L2 where Strevens's (1980) study was quoted to emphasize that L2 students who cannot master non science-specific words will encounter problems learning science through English. Similarly, Clark's (1997) and Prophet & Towse's (1999) work also show EMI science students struggle with the general English that is typically known by L1 students, such as general vocabulary and connectives to express relationships.

In summary, when EMI students report problems with technical vocabulary, the cause could be the language contained in the explanation or definition of the technical concept as opposed to the form or surface literal meaning of the technical word. However, having pointed out the strong need of attention to non-technical vocabulary in EMI contexts, this is not to say that non-technical vocabulary or everyday meaning of technical vocabulary does not need to be explained at all in L1 classrooms. It has been shown that general academic vocabulary could also cause comprehension problem for L1 students (Prophet & Towse, 1999). Rather, the argument is that the typically common vocabulary known to L1 students may be challenging to EMI students. To further examine the kind of language instruction provided in EMI and non-EMI contexts, such as the type of linguistic features explained and how they are explained, research that compares monolingual teachers in these two contexts is needed.

While previous studies provided little empirical evidence quantifying the type of language features addressed in EMI secondary school classes, the qualitative description in Lo's (2010) study also reports the bilingual EMI teachers' attention to the non-technical vocabulary and phrases in EMI secondary schools in Hong Kong (e.g., the meaning of *glow*, the use of *towards each other*). Although not quantified, this finding resonates with the monolingual teachers' attention to non-technical vocabulary in this current study. However, the quantitative results available at the tertiary level indicate a primary focus on the meaning of technical vocabulary and limited attention to other linguistic features, e.g., pronunciation, spelling, grammatical features, and attention to non-technical vocabulary was also rare (Basturkmen & Shackleford, 2015; Costa, 2012). This difference could be partly due to the level of educational phase, i.e., university vs secondary. It could also be reflective of the monolingual science teachers' tendency to use a wide range of non-technical everyday vocabulary and colloquial expressions in their lessons. As Llurda (2005) summarized, 'native-speaker' language teachers are often seen as possessing a more skilful mastery of a wide range of vocabulary. This characteristic is to a degree also reflected in the type of LFEs provided by monolingual science teachers in this EMI context.

Among the monolingual teachers' instruction on the informal and colloquial use of English, the presence, although minimal, of explaining idioms also reflects the variety of English that is used and taught in these EMI classes. This touches upon the topic of Global Englishes (Galloway & Rose, 2015), an area attracting growing interests from applied linguists. Taking the notion of three circles of Englishes established by Kachru (1984), the students in this study, being in the expanding circle which is made up of countries without a history of Anglo-American colonization, were taught through the inner circle varieties of English. Extract 7 demonstrates that in this EMI scenario, with monolingual teachers from a specific anglophone region, the variety of English, i.e., the 'E' in the 'EMI', is also reflected in the LFEs provided.

Regarding grammatical LFEs, these were relatively few and only featured a focus on morphemes. As highlighted by many scholars, in communication-oriented classes overt instruction on a variety of grammatical features is needed to improve accuracy in students' receptive language skills (Swain, 1996) and to develop a proper level of academic literacy (De Oliveira & Schleppegrell, 2015). In comparison with Lo's (2010) work with bilingual EMI teachers in Hong Kong where explicit explanations of verb tenses and part of speech were found, in this study there was an absence of them. This may be indicative of the monolingual teachers' lack of awareness and attention to a wide variety of grammatical features that their students may have difficulty with. Again, this behaviour also reflects what was discussed in the literature on NS and NNS language teachers where NS teachers are often commented on as lacking in grammar and linguistic awareness (Llurda, 2005; Moussu & Llurda, 2008).

What was also absent in the LFEs of this study is the language structures that are distinctive to the register of science, such as connectives to describe logical relationships, complex noun phrases and nominalization (Halliday, 1993) and so on. Built upon the perspective of systemic functional linguistics, these grammatical features have been widely argued as crucial for building academic literacy particularly at the secondary school level. Although they also need to be systematically and explicitly instructed for L1 (De Oliveira & Schleppegrell, 2015; Gibbons, 2002), they may need even greater focus for EMI students. Similar to this study, explicit teaching of these features was also not commonly found in bilingual EMI teachers' classes (Basturkmen & Shackleford, 2015; Costa, 2012; Lo, 2010; Pérez-Vidal, 2007; Walker, 2011).

5.3 L1 vs L2

As mentioned earlier, most of the existing EMI studies, including the ones that examined classroom practices, feature bilingual teachers who share students' L1. The monolingual teacher sample in this study determined that the LFEs were conducted through L2-only.

Bilingual EMI teachers have been observed to use L1 in various ways to facilitate content teaching, such as explaining difficult vocabulary, describing daily life experience in relation to the subject content and providing emphasis (Lin, 1996; Probyn et al., 2002; Lo, 2010), and scaffolding students' output in L1 and developing it in L2 (Lin & Wu, 2015; Urmeneta & Walsh, 2017). In language teaching episodes in the EMI classes of previous studies, L1 was also observed to explain various language aspects, check students' understanding and provide direct translations (Costa, 2012; Lo, 2010; Maxwell-reid, 2017). It has also been shown in language classes that teachers' understanding of students' L1 can allow them to provide the English equivalents, which is shown to be beneficial for students' L2 learning (Macaro, Nakatani, Hayashi, & Khabbazbashi, 2014). Bilingual teachers are also argued to have the advantage of choosing strategically how to use each language in giving explanations, such as explaining by giving definitions, paraphrasing, giving examples and so on (Macaro, Guo, Chen, & Tian, 2009). However, while L1 could play all these roles, the monolingual teachers in this study cannot.

Because of the monolingual teacher population of this study, the instances of vocabulary-focused LFEs where a Chinese equivalent exists but instruction had to be all through English provided some interesting insights on the issue of L1 use. As Extract 2 showed, the teacher was not able to point out the Chinese equivalent of the target word *anchor*, nor would she know that it was equivalent both in its original and its figurative meaning. It could be argued that the L1 translation is a more efficient way of clarifying the meaning of the word (and, indeed, then recalling it, see Zhao & Macaro, (2016)), as compared to the two-minute-and-a-half explanation in L2. This is an important factor to consider in EMI contexts where a balance between the learning of content and language needs to be arrived at. However, a counter argument could be that lengthy L2-only explanation provided students with more exposure to L2, particularly the non-academic kind, which the students probably are rarely exposed to outside of the school context. In Extract 2, the teacher also used a variety of non-academic lexical items that are associated with *anchor*, such as *hold, in one spot, sink, float away, blow away* and so on. This rich description provides ample input which could be comprehensible to students with detailed contextual information of how boats are anchored, thus creating one beneficial condition for second language acquisition to take place, i.e., comprehensible input (Krashen, 1982), and may also serve to develop students' L2 inferencing strategies. Opportunities of negotiation of meaning where students need to comprehend teachers' input in L2 to understand the content knowledge is also a reason why EMI classes are believed to provide a more favorable environment for L2 learning than language classes.

In the background to the above discussion is the recent line of work on translanguaging (García & Li, 2014) which suggests more recognition be given of the benefit of integrating students' L1 at the conceptualization level and classroom practice level. In EMI contexts where L2 English students face difficulties not only in understanding technical vocabulary but also very often the non-technical ones, as the findings show, the L1 equivalent could be helpful in efficiently clarifying the meaning of a target word, which is important considering the potentially high quantity of unknown vocabulary there is for EMI students. At the same time, L2-only explanation with ample details could provide students the much-needed exposure to non-academic English. As Joyce (2018) proposed, *when the goal is primarily vocabulary expansion, the L1 method is preferable. On the other hand, for the purposes of general language development, learning through an L2 definition is favoured (p. 225).*

6. Conclusion

This study has taken a perspective different from the majority of the previous EMI studies by examining data from monolingual science teachers' classes. An obvious limitation is that it has only examined the interactional practices of a small and very specific number of such teachers. It should also be noted that the findings need to be understood within the context of the foreign high school programs in China, e.g., students' motivation of going overseas for university education, the use of curricula designed for L1 students, the foreign teachers' background and so on, and we have not provided here, because of space limitations, the students' perspective nor that of the teachers. However, the examination of LFEs in this study provides a piece of the jigsaw in our understanding of the extent and kind of explicit instruction of language that takes place in EMI classes.

The implication for teaching is that explicit instruction on language needs to be given more attention, and awareness needs to be developed regarding how the non-technical vocabulary in science explanations can cause problems. Attention to a wider range of grammatical features probably needs to be developed by monolingual science teachers, together with how language is used beyond individual vocabulary to express meaning in science texts. This is especially the case if parallel English language instruction does not provide that kind of focus. Lastly, the use of students' L1 in EMI contexts and the efficiency it provides need to be continually reconsidered and balanced against the exposure to the L2, particularly to non-academic language, which may be rare for EMI students.

References

Basturkmen, H., & Shackleford, N. (2015). How content lecturers help students with language: An observational study of language-related episodes in interaction in first year accounting classrooms. *English for Specific Purposes*, 37(1), 87–97. https://doi.org/10.1016/j.esp.2014.08.001

Butzkamm, W. (1998). Code-switching in a bilingual history lesson: The mother tongue as a conversational lubricant. *International Journal of Bilingual Education and Bilingualism*, 1(2), 81–99. https://doi.org/10.1080/13670059808667676

Cammarata, L., & Tedick, D. J. (2012). Balancing content and language in instruction: The experience of immersion teachers. *The Modern Language Journal*, 96(2), 251–269. https://doi.org/10.1111/j.1540-4781.2012.01330.x

Clark, J. (1997). Beyond the turgid soil of science prose: STAP's attempt to write more accessible science text materials in general science. In M. Sanders (Ed.), *Proceedings of the Fifth Annual Meeting of the Southern African Association for Research in Science and Mathematics Education*. (pp. 390–396). Johannesburg, South Africa: University of the Witwatersrand.

Coleman, J. (2006). English-medium teaching in European higher education. *Language Teaching*, 39(1), 1–14. https://doi.org/10.1017/S026144480600320X

Cook, V. (1999). Going beyond the native speaker in language teaching. *TESOL Quarterly*, 33(2), 185–209. https://doi.org/10.2307/3587717

Costa, F. (2012). Focus on form in ICLHE lectures in Italy: Evidence from English-medium science lectures by native speakers of Italian. *AILA Review*, 25(1), 30–47. https://doi.org/10.1075/aila.25.03cos

Coxhead, A. (2000). A new academic word list link. *Tesol Quarterly*, 34(2), 213–238. https://doi.org/10.2307/3587951

De Graaff, R., Koopman, G. J., Anikina, Y., & Westhoff, G. (2007). An observation tool for effective L2 pedagogy in Content and Language Integrated Learning (CLIL). *International Journal of Bilingual Education and Bilingualism*, 10(5), 603–624. https://doi.org/10.2167/beb462.0

De Oliveira, L. C., & Schleppegrell, M. (2015). *Focus on grammar and meaning*. Oxford: Oxford University Press.

Ellis, R., & He, X. (1999). The roles of modified input and output in the incidental acquisition of word meanings. *Studies in Second Language Acquisition*, 21(2), 285–301. https://doi.org/10.1017/S0272263199002077

Fang, Z. (2005). Scientific literacy: A systemic functional linguistics perspective. *Science Education*, 89(2), 335–347. https://doi.org/10.1002/sce.20050

Galloway, N., & Rose, H. (2015). *Introducing global Englishes*. London: Routledge. https://doi.org/10.4324/9781315734347

García Mayo, M., & Basterrechea, M. (2017). CLIL and SLA: Insights from an interactionist perspective. In A. Llinares & T. Morton (Eds.), *Applied linguistics perspectives on CLIL* (pp. 33–50). Amsterdam: John Benjamins. https://doi.org/10.1075/lllt.47.03gra

García, O., & Li, W. (2014). *Translanguaging: Language, bilingualism and education.* Houndmills: Palgrave. https://doi.org/10.1057/9781137385765

Gibbons, P. (2002). *Scaffolding language, scaffolding learning: Teaching second language learners in the mainstream classroom.* Portsmouth, NH: Heinemann.

Gorard, S., & Taylor, C. (2004). *Combining methods in educational and social research.* Maidenhead: Open University Press.

Halliday, M. A. K. (1993). Towards a language-based theory of learning. *Linguistics and Education,* 5(2), 93–116. https://doi.org/10.1016/0898-5898(93)90026-7

Joyce, P. (2018). L2 vocabulary learning and testing: the use of L1 translation versus L2 definition. *Language Learning Journal,* 46(3), 217–227. https://doi.org/10.1080/09571736.2015.1028088

Kachru, B. B. (1984). World englishes and the teaching of english to non-native speakers, contexts, attitudes, and concerns. *TESOL Newsletter,* 18, 25–26.

Kachru, B. B. (1985). Standards, codification, and sociolinguistic realm: the English language in the outer circle. In R. Quirk & H. G. Widdowson (Eds.), *English in the world: Teaching and learning the language and literatures* (pp. 11–30). Cambridge: Cambridge University Press.

Krashen, S. (1982). *Principles and practice in second language acquisition.* New York, NY: Prentice Hall.

Lasagabaster, D., & Huguet, A. (2007). *Multilingualism in European bilingual contexts: Language use and attitudes.* Clevedon: Multilingual Matters.

Laufer, B. (2005). Focus on form in second language vocabulary learning. In S. Foster-Cohen, M. Garcia-Mayo, & J. Cenoz (Eds.), *Eurosla Yearbook: Volume 5* (pp. 223–250). Amsterdam: John Benjamins.

Lemke, J. L. (1990). *Talking science: Language, learning and values.* Norwood, NJ: Ablex.

Lin, A. M. Y. (1996). Bilingualism or linguistic segregation? Symbolic domination, resistance and codeswitching in Hong Kong Schools. *Linguistics and Education,* 8(1), 49–84. https://doi.org/10.1016/S0898-5898(96)90006-6

Lin, A. M. Y., & Wu, Y. (2015). 'May I speak Cantonese?' – Co-constructing a scientific proof in an EFL junior secondary science classroom. *International Journal of Bilingual Education and Bilingualism,* 18(3), 289–305. https://doi.org/10.1080/13670050.2014.988113

Llinares, A., & Lyster, R. (2014). The influence of context on patterns of corrective feedback and learner uptake: A comparison of CLIL and immersion classrooms. *The Language Learning Journal,* 42(2), 181–194. https://doi.org/10.1080/09571736.2014.889509

Llurda, E. (Ed.). (2005). *Non-native language teachers: Perceptions, challenges, and contributions to the profession.* Boston, MA: Springer. https://doi.org/10.1007/b106233

Lo, Y. Y. (2010). What happens to classroom interaction patterns and teachers' code-switching behaviour when the medium of instruction changes?: An exploratory study in Hong Kong secondary schools (Unpublished dotoral dissertation). University of Oxford.

Long, M. (1996). The role of the linguistic environment in second language acquisition. In W. C. Ritchie & T. K. Bhatia (Eds.), *Handbook of second language acquisition* (pp. 413–468). New York, NY: Academic Press.

Lorenzo, F. (2007). An analytical framework of language integration in L2 content-based courses: The European dimension. *Language and Education*, 21(6), 502–514. https://doi.org/10.2167/le708.0

Lyster, R. (2007). *Learning and teaching languages through content: A counterbalanced approach*. Amsterdam: John Benjamins. https://doi.org/10.1075/lllt.18

Macaro, E. (2018). *English medium instruction: Content and language in policy and practice*. Oxford: Oxford University Press. https://doi.org/10.30687/978-88-6969-227-7/001

Macaro, E., Curle, S., Pun, J., An, J., & Dearden, J. (2018). A systematic review of English medium instruction in higher education. *Language Teaching*, 51(1), 36–76. https://doi.org/10.1017/S0261444817000350

Macaro, E., Guo, T., Chen, H., & Tian, L. (2009). Can differential processing of L2 vocabulary inform the debate on teacher codeswitching behaviour: The case of Chinese learners of English. In B. Richards, M. Daller, P. Malvern, J. Meara, J. Milton, & J. Treffers-Daller (Eds.), *Vocabulary studies in first and second language acquisition: The interface between theory and application*. (pp. 125–146). London: Palgrave Macmillan. https://doi.org/10.1057/9780230242258_8

Macaro, E., Nakatani, Y., Hayashi, Y., & Khabbazbashi, N. (2014). Exploring the value of bilingual language assistants with Japanese English as a foreign language learners. *The Language Learning Journal*, 42(1), 41–54. https://doi.org/10.1080/09571736.2012.678275

Marsh, D. (2002). *CLIL/EMILE – The European dimension: Actions, trends and foresight*. Potential Public Services Contract DG EAC.

Martin, J. (1991). Norminalization in science and humanities: Disliking knowledge and scaffolding text. In E. Ventola (Ed.), *Functional and systemic linguistics: Approaches and uses* (pp. 307–337). Berlin: Mouton de Gruyter. https://doi.org/10.1515/9783110883527.307

Maxwell-reid, C. (2017). Classroom discourse in bilingual secondary science: Language as medium or language as dialectic? *International Journal of Bilingual Education and Bilingualism*, 1–14. https://doi.org/10.1080/13670050.2017.1377683

McMillan, B., & Turnbull, M. (2009). Teachers' use of the first language in French immersion: Revisiting a core principle. In M. Turnbull & J. Dailey-O'Cain (Eds.), *First language use in second and foreign language learning* (pp. 15–34). Clevedon: Multilingual Matters. https://doi.org/10.21832/9781847691972-004

Met, M. (1999). *Content-based instruction: Defining terms, making decisions*. Washington D.C: The National Foreign Language center.

Miller, J. (2009). Teaching refugee learners with interrupted education in science: Vocabulary, literacy and pedagogy. *International Journal of Science Education*, 31(4), 571–592. https://doi.org/10.1080/09500690701744611

Moussu, L., & Llurda, E. (2008). Non-native English-speaking English language teachers: History and research. *Language Teaching*, 41(03), 315–348. https://doi.org/10.1017/S0261444808005028

Nikula, T., Dalton-Puffer, C., & García, A. L. (2013). CLIL classroom discourse: Research from Europe. *Journal of Immersion and Content-Based Language Education*, 1(1), 70–100. https://doi.org/10.1075/jicb.1.1.04nik

Norris, J.M., & Ortega, L. (2000). Effectiveness of L2 instruction: A research synthesis and quantitative meta-analysis. *Language Learning*, 50(3), 417–528. https://doi.org/10.1111/0023-8333.00136

Pérez-Cañado, M. L. (2012). CLIL research in Europe: Past, present, and future. *International Journal of Bilingual Education and Bilingualism*, 15(3), 315–341. https://doi.org/10.1080/13670050.2011.630064

Pérez-Vidal, C. (2007). The need for focus on form (fof) in content and language integrated approaches: An exploratory study. *Resla*, 1, 39–54.

Pica, T., Young, R., & Doughty, C. (1987). The impact of interaction on comprehension. *Tesol Quarterly*, 21, 737–758. https://doi.org/10.2307/3586992

Probyn, M., Murray, S., Bothal, L., Botya, P., Brooks, M., & Westphal, V. (2002). Minding the gaps – An investigation into language policy and practice in four Eastern Cape town districts. *Perspectives in Education*, 20(1), 29–46.

Prophet, B., & Towse, P. (1999). Pupils' understanding of some non-technical words in science. *School Science Review*, 295, 79–86.

Robson, C. (2002). *Real world research: A resource for social scientists and practitioner-researchers* (2nd ed.). Oxford: Blackwell.

Rollnick, M. (2000). Current issues and perspectives on second language learning of science. *Studies in Science Education*, 35(1), 93–121. https://doi.org/10.1080/03057260008560156

Seah, L. H., Clarke, D. J., & Hart, C. E. (2014). Understanding the language demands on science students from an integrated science and language perspective. *International Journal of Science Education*, 36(6), 952–973. https://doi.org/10.1080/09500693.2013.832003

Shohamy, E. G. (2006). *Language policy: Hidden agendas and new approaches*. London: Routledge. https://doi.org/10.4324/9780203387962

Spada, N., & Tomita, Y. (2010). Interactions between type of instruction and type of language feature: A meta-analysis. *Language Learning*, 60(2), 263–308. https://doi.org/10.1111/j.1467-9922.2010.00562.x

Strevens, P. (1980). *Teaching English as an international language: From practice to principle* (1st ed.). Oxford: Pergamon Press.

Swain, M. (1996). Integrating language and content in immersion classrooms: Research perspectives. *Canadian Modern Language Review*, 52, 529–548. https://doi.org/10.3138/cmlr.52.4.529

Urmeneta, C. E., & Walsh, S. (2017). Classroom interactional competence in content and language integrated learning. In A. Llinares & T. Morton (Eds.), *Applied linguistics perspectives on CLIL* (pp. 183–200). Amsterdam: John Benjamins. https://doi.org/10.1075/lllt.47.11esc

Walker, E. (2011). How 'language-aware' are lesson studies in an East Asian high school context? *Language and Education*, 25(3), 187–202. https://doi.org/10.1080/09500782.2011.555557

Williams, J., & Doughty, C. (1998). *Focus on form in classroom second language acquisition*. Cambridge: Cambridge University Press.

Yip, D. Y., Tsang, W. K., & Cheung, S. P. (2003). Evaluation of the effects of medium of instruction on the science learning of Hong Kong secondary students: Performance on the science achievement test. *Bilingual Research Journal*, 27(2), 295–331. https://doi.org/10.1080/15235882.2003.10162808

Zhao, T., & Macaro, E. (2016). What works better for the learning of concrete and abstract words: Teachers' L1 use or L2-only explanations? *International Journal of Applied Linguistics*, 26(1), 75–98. https://doi.org/10.1111/ijal.12080

Appendix

A preview of the background of the 15 teachers

Province	School	Curriculum	T	Subject	Gender	Age	Nationality
Province A	Sch 1	Canadian British Columbia	1	Chemistry	F	33	Canadian
			2	Physics	M	54	Canadian
			3	Biology	F	52	Canadian
	Sch 2	British IGCSE, AS, A2	4	Biology	M	29	American
	Sch 3	Canadian British Columbia	5	Physics	M	25	Canadian
			6	Chemistry	M	59	Canadian
			7	Biology	F	24	Canadian
	Sch 4	Canadian Alberta	8	Physics	M	56	Canadian
Province B	Sch 5	American AP program	9	Biology	M	34	American
	Sch 6	IB program	10	Biology	M	36	American
Province C	Sch 7	Canadian British Columbia	11	Physics	M	24	Canadian
			12	Chemistry	F	23	Canadian
			13	Biology/Geology	F	31	Canadian
			14	Biology	F	29	Canadian
		British IGCSE, AS, A2	15	Biology	M	32	British

The positioning of Japanese in a secondary CLIL science classroom in Australia
Language use and the learning of content

Marianne Turner
Monash University

In Australia, content and language integrated learning (CLIL) is commonly implemented as a way to encourage innovation in language teaching. This paper explores how Japanese can also be used to innovate the teaching of content. Qualitative data are drawn from a Year 8 science Japanese CLIL classroom in a secondary school with an opt-in CLIL program. In the class, a monolingual (in English) science teacher was co-teaching with a Japanese language teacher. Findings from observations, after-class reflections, teacher and student interviews, a student survey and work samples revealed that students were highly engaged with the Japanese component of their science lessons. Kanji was further positioned as a way for students to deepen their understanding of scientific concepts. However, there also appeared to be a separation in the way both teachers and students spoke about Japanese language use and learning science. Implications of these findings are discussed in the paper.

Keywords: CLIL, Japanese, secondary education, science

1. Introduction

In Europe, content and language integrated learning (CLIL) is an umbrella term that has been conceptualized in relation both to methodology (e.g., Coyle, 2007; Coyle, Hood, & Marsh, 2010; Eurydice, 2006) and to common characteristics in its implementation (Dalton-Puffer, Llinares, Lorenzo, & Nikula, 2014). These European characteristics include the CLIL language being a *lingua franca*, the CLIL classroom as the domain of the subject area teacher, and language classrooms operating independently of the CLIL classroom (ibid.) Explicit language management, such as 'clear separation of teacher use of one language versus another for sustained periods of time' (Fortune & Tedick, 2008, p. 10), has not been a key feature of CLIL. This is particularly apparent in Czura and Papaja's (2013) study on CLIL programs in Poland. They identified four different models of delivery, all based on the amount of target language in the CLIL classrooms: 80:20 percent (in favour of the target language), 50:50, 10–50 percent of the target language, and a model that included strategies using the target language sporadically or in written form. Mapping gains in students' language proficiency has also not always been the goal of teachers. For example, in their discussion of the Austrian context, Hüttner, Dalton-Puffer and Smit (2013, p. 280) found that success was constructed as affective. As they said, 'one might argue that stakeholders can make CLIL into a success more easily by changing their or their learners' feelings as [target language] speakers, than by changing the language proficiency of a large and mixed cohort of learners'.

This flexible approach to cross-curricular language use is well-established in Australia (see Fielding & Harbon, 2015; Smala, 2016) and can be related as much to methodology as to the way a program is structured (Cross, 2015). For example, in the state of Victoria, the Government endorses training in CLIL that focuses on the pedagogical framework known as the 'four Cs' – content (subject matter), communication (language), cognition (thinking and learning skills), and culture/citizenship (social awareness of self and others) (see Coyle, 2007, p. 550). In their evaluation of CLIL programs in Victoria, Cross and Gearon (2013) found that school support, but not necessarily school commitment, was an important factor. Their findings indicated that it was viable for teachers to initiate CLIL programs without schools needing to make any changes to organization or staffing. Outside twelve Government-funded bilingual education programs, in which a funding incentive was recently put in place to move to teaching 50 percent of the time in the target language, there are no specifications around how much instructional time in a target language is required in order to implement CLIL (Department of Education and Training, Victoria, 2017):[1] schools and teachers can choose the extent to which they include the target language in class. In this article, I will explore how choices that related to the use of both a CLIL language and English positively influenced students' conceptual understanding of secondary school science but also appeared to create a separation between communicative target language use and the subject area.

The CLIL language under discussion is Japanese – a language that has been taught in Australia for over three decades (de Kretser & Brown, 2010). Japanese teachers were among the first to trial CLIL in Victoria (see Cross & Gearon,

---

1. In these programs, 50% of instruction over a week is expected to be delivered in the target language. Program funding is tied to this, and the stipulation of language use is considered to be a form of external language management.

2013), and have experimented with the approach in different ways, both inside and outside the language classroom (Turner, 2013). Similar to other CLIL languages, Japanese CLIL has mostly been implemented in the junior years in secondary schools. Engaging students in meaningful language use and, indirectly, increasing retention in senior secondary has been an imperative. Student engagement with languages at school is relatively low; in the final two years, languages are an elective, and there is a large attrition rate. For example, in 2015, only 6.9% of students in Victorian Government schools elected to study a language (Department of Education and Training, Victoria, 2016). I will now review research relevant to content and language integration in the area of science, the content area under discussion in this article, and language specific issues related to the teaching of content in immersion settings, before reporting on findings from a Japanese CLIL junior secondary science classroom in Victoria, Australia.

2. Literature review

2.1 Science and content and language integration

There is a general consensus in the research that has been conducted on CLIL programs that CLIL is beneficial for students' language proficiency (Pérez-Cañado, 2012; Pérez-Vidal & Roquet, 2015), and science has been a popular choice of subject for CLIL (e.g., Fernanández-Sanjurjo, Fernández-Costales, & Arias Blanco, 2017; Rasulo, de Meo, & de Santo, 2017). In light of this, attention has been paid to the integration of language and science within a broader focus on different subject areas. For example, Dalton-Puffer developed the idea of cognitive discourse functions, or 'the patterns which emerge from the needs humans have when they deal with cognitive content for the purposes of learning, representing and exchanging knowledge' (2016, p. 31). Dalton-Puffer's (2013, 2016) work developed the Council of Europe's (2013) systematic cataloguing of the English language required in science, mathematics, history and literature in lower secondary classrooms. This approach to the integration of language structures needed by students to learn content is mirrored in literature on assisting English-as-an-additional-language (EAL) learners in science classrooms in English-speaking countries. For example, Oliveira (2017) discussed the application of a meaning-based theory of language – systemic functional linguistics (see Halliday & Matthiessen, 2004) – to science classrooms in the United States.

Language-sensitive strategies for integrating language with content in the science classroom have therefore been developed across EAL and CLIL contexts, and include 'front-loading' language before teaching the content (Bravo, 2017), collaborating with language teachers (Rasulo et al., 2017), and paying attention to discourse at different stages of the science lesson (Nikula, 2015). These strategies can be very practical and classroom-focused. For example, Silva, Weinburgh, Malloy, Horak Smith & Nettles Marshall (2012) demonstrated how a '5R' instructional model, developed by Weinburgh and Silva (2010) for the science classroom, could be used. The instructional model focused on the importance of language, and the 5Rs referred to replace (everyday language with scientific language), reveal (explain scientific language with no corresponding every day word), reposition (move students from familiar language to the relevant scientific concept), repeat (use important language multiple times during explanations), and reload (recycle language in later lessons).

Although language has been a strong focus of research in CLIL and content-based science classrooms, content knowledge has been less of a focus (Cenoz, Genesee, & Gorter, 2014; Paran, 2013). There are studies in which students studying in a CLIL program were found to perform better in the subject area than their non-CLIL peers (e.g., Ullman, 1999; Wode, 1999), but a recent study conducted in Spain showed that primary students studying science in English performed slightly worse than their non-CLIL peers, and that lower socio-economic status was also linked to lower scores (Fernández-Sanjurjo, Fernández-Costales, & Arias Blanco, 2017). The improvement of students' cognitive abilities has also been under-researched, although Jäppinen's (2005) study is a notable exception. In this longitudinal study of 669 Finnish learners of science and mathematics, Jäppinen investigated the stimulation of students' processes of cognition in a CLIL program. The five critical discovery learning areas that Jäppinen (2005, p. 153–154) used in the study were:

1. awareness of existing concepts and the ability to understand, use, apply, and explain them,
2. awareness of meaning schemes and the ability to create links between them,
3. the ability to exploit information in problem-solving situations and to formulate hypotheses,
4. the ability to solve problems that are difficult to explain to oneself or when concepts that one has not yet acquired are involved, and
5. the ability to exploit the flow of information, to make comparisons and form antitheses, and to choose between two or more alternatives.

These critical discovery learning areas link with both understanding and the displaying of this understanding, especially the first two areas. Cognition is a significant aspect of a subject area with a high level of conceptual demand, such as science. Recently, CLIL literature has also had a more holistic focus; rather than exploring language, content and cognition as separate aspects of CLIL, the integration dimension is highlighted (Meyer, Coyle, Halbach, Schuck, & Ting, 2015; Nikula, Dalton-Puffer, Llinares, & Lorenzo, 2016). This focus prioritises literacies intrinsic to different discipline areas (such as science) and integrated curricula.

2.2 Language-specific issues in using a target language to teach content

In many contexts where science is taught as part of a CLIL or content-based program, English is the target language of choice, and this is reflected in the literature. However, if we look at contexts where attention is paid to language management and widen the lens to content in general, other languages become more visible in (Anglophone) research. In both two-way immersion programs in the United States and French immersion programs in Canada, research has focused on language proficiency in the target language and English (e.g., De Jong, 2002; Oller & Eilers, 2002; Tedick & Young, 2014; Thomas & Collier, 2002; Swain, 1985, 2000), ways language is used in classroom interactions (e.g., Lyster, 2007; Lyster & Mori, 2006) and on academic achievement (Genesee, 1987, 2004; Lindholm-Leary, 2001; Lindholm-Leary & Genesee, 2010; Martin-Beltrán, 2010).

Although the relationship between different languages and particular content areas has not yet been a strong focus in the literature, Lee and Lee (2017) argued for a language-specific pedagogical knowledge, as well as a language-specific content curricula, in their context of Korean/English two-way immersion programs in the United States. Lee and Lee (2017, p. 2) maintained that pedagogical language knowledge (PLK) includes aspects of pedagogical content knowledge and understood that 'teachers who have a strong PLK basis would have a solid understanding of which aspects of their subject are typically easy for students and which are typically more difficult, including knowledge about how idiosyncratic features of the language of instruction can play a role in the teaching and learning process'. In their study, based in the subject area of mathematics, they found that the Korean teachers struggled with PLK and would have benefited from support in this area. Lee and Lee (2017) considered there to be a need for a language-specific curriculum that took into account cultural and linguistic characteristics of Korean.

The study discussed in this article reinforces this need to take into account the significance of cultural and linguistic characteristics of a language – Japanese. In Japan, the writing system is complex in that it involves three different scripts – *hiragana* (syllabic writing usually used for inflections and function words), *katakana* (syllabic writing usually used for foreign words), and *kanji* (Chinese characters). In Japan, this complexity is referred to as *kokuji mondai*, or the national script problem (Hovhannisyan, 2018). The way Japanese language should be written became a subject of debate during the modernisation of Japan in the Meiji period (Hovhannisyan, 2018) and continues in the modern era (e.g., Abe,

2002, 2015; Tanaka, 2011). One important reason given for simplification of the writing system is the acknowledgement of the challenges involved in learning three extra scripts for people who are learning Japanese as an additional language[2] (Abe, 2015; Andō, 1942). In CLIL programs, the distribution of focus on oral communication and the complex writing system can be left to the teachers to navigate. The way teachers chose to navigate the issue resulted in the findings reported upon in this article; namely, the conceptual advantages of using both *kanji* and English in the science classroom, and the apparent separation between communicating in the target language (Japanese) and the subject area.

3. The study

The data discussed in the article were drawn from a larger qualitative study on opportunities, engagement and outcomes related to CLIL in one secondary school in Australia. The participating school was a Catholic, co-education school with just under 2,000 students. There was a school commitment in that CLIL classes were structured into the curriculum for Years 7–9 students who had opted in to the program. The number of students in the program was quite low (see Table 1). These students were grouped according to their year level and studied with the same group over the three years. Teachers had access to CLIL training. There were no selection criteria for the two CLIL programs (Japanese and Italian). Both programs taught humanities and religious education in the CLIL language and, as a result of the availability of Japanese-speaking staff, science and visual arts were also taught in Japanese. Students' Japanese language proficiency was beginner level. Some students had studied Japanese in a language classroom in primary school but this was not a prerequisite. No students had a Japanese heritage. Teachers could choose how they incorporated Japanese into their subject area, and the extent of the use varied considerably between classes.

Table 1. Number of students at each year level studying CLIL

Year level	Japanese CLIL	Italian CLIL
Year 7	0	7
Year 8	13	15
Year 9	8	12

---

2. Chinese students of Japanese will already know the way of writing the *kanji* (Chinese characters) but, for other learners, all three scripts would be additional.

Results in this article are drawn from the Year 8 science class. In this class, two teachers – Audrey and Louise (both pseudonyms) – were co-teaching. Audrey was employed as a science teacher and taught science to all the Year 8s, including the CLIL class. Audrey identified as being monolingual in English. Louise was employed as a Japanese support officer for all the Japanese CLIL classes but had a co-teaching role in the science classes. Louise had studied science at university and was also transitioning to be the sole science CLIL teacher. She was a qualified teacher and had learned Japanese as a foreign language.

Table 2. Teacher participants

Teacher	CLIL language	Content area in study	Teaching qualifications	CLIL teaching experience in context of overall experience
Audrey	Japanese (non-Japanese speaker)	Science	Science	Science – 1st year; co-teaching with Louise; At least 6 years teaching experience.
Louise	Japanese	Science	Japanese (+ Bachelor of Science)	Science – 1st year; Co-teaching with Audrey; All other Japanese CLIL classes: support; 1 year teaching experience.

Data from the Year 8 science class were collected from two video-recorded classroom observations, joint teacher reflections after each of the two lessons (stimulated by a video-recording of the lesson), separate teacher interviews, six student interviews, lesson plans and resources, a student survey and student work samples from the observed classes. Teachers and students were interviewed (and students were surveyed) about their general perspectives on, and experiences with, CLIL; for example, participants were asked how they felt about CLIL (including its strengths and challenges), their perceptions around language of assessment, and around differences between the CLIL content area classroom and the language classroom. Surveys further included questions on satisfaction with language use in CLIL classes, as well as understanding of the CLIL language. All interviews and reflections were transcribed. Data were analysed using thematic analysis (Braun & Clarke, 2006). Themes were identified by cross-referencing data; for example, both teachers and students spoke of *kanji* assisting conceptual understanding of science, and the resources used to promote this understanding were sighted. Other instances of this cross-referencing will be used as support in the findings section below, where themes are discussed.

4. Findings

Two factors in the science classroom were found to be significant in the study. First, the science class was content driven and Japanese language use was conceptualised as a bonus to the learning of content. Second, Louise chose to use *kanji* to reinforce scientific concepts students were learning in class. This decision gave Japanese a conspicuous role in the learning of content. However, except for the *kanji*, which appeared to occupy an intermediary space, Japanese language use was found to be conceptualised as distinct from the learning of science.

4.1 Using kanji to reinforce scientific concepts

A significant finding in the data was the prevalence of *kanji* in teaching and learning. Louise's focus was found to be on helping the students understand scientific concepts where the meaning appeared to be easier to access in *kanji* than in English. Tasked with bringing Japanese into the science curriculum, Louise had noticed this benefit of *kanji* and thus did not choose only to write words in *hiragana*. Even though *hiragana* was an easier script for the students to decode, she did not understand it to have the same conceptual advantage of *kanji*. She explained her choice in interview:

> I tend to use a lot of kanji because they're characters that contain meaning, like they're not just phonetic they have the meaning as well. [...] I've noticed that benefit, and so I do try to teach the words and the kanji, you know put a lot of kanji on the slide shows, or well just you know, the key words, on the slides shows mixed in with the English. (Louise, interview)

As is evident in the slide show sample below, Louise included the *kanji* to supplement the English, and this supplementary material could then be used to help students understand the concept because the literal translation came back into English in a different way – 'hard body' for 'solid', for example (see Figure 1). Literal translations could then be reinforced in scientific explanations (see Figure 2). The emphasis appeared to be on harnessing Japanese to assist in the learning of subject matter rather than on immersion in Japanese. The approach allowed for the introduction of relatively complex *kanji* because there was a content-based reason for learning it and it was directly connected to its English equivalent.

In interview, Louise reported that she herself found *kanji* to be useful for the understanding of scientific concepts, and this may have been a reason for her choice of approach:

Figure 1. Slide show sample 1: Using kanji to help explain scientific concepts

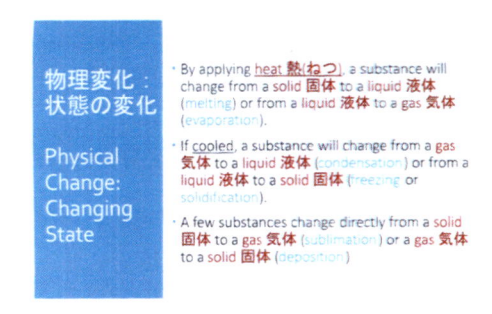

Figure 2. Slide show sample 2: Using kanji to help explain scientific concepts

> [Kanji can help students understand the scientific concepts] because it's very easy to see the literal translation of what the words mean. [...] Like with [...] fertilisation, so the character for 'receive' and the character for 'sperm', receive sperm, that's fertilisation. [...] You know, just things like can help [the students] see. Variable, the word variable, [...] when you're designing an experiment, you've got to design your variables, [...] the kanji for that is 'change', 'number'. So even for me, that helped me understand the word variable better.　　　(Louise, interview)

One of the students, Rachel, also highlighted this conceptual advantage of *kanji* when she was asked in interview if she thought that CLIL was helping her learn different subject areas. The other students understood CLIL to be helping them learn different subject areas but mainly chose to talk about this in a general way, or by saying that they liked Japanese more than they liked the subject area. Rachel chose to speak about the science CLIL classroom in a specific way:

> I think [I know more about] science cause when we look at science, there are [...] different words to put together and it makes the definition, so in a way the kanji itself is giving the definition of the word. So, for example, exothermic, we have

> the word heat, like thermic for heat, and we have another word and then when we put it together it's technically just giving the definition of the word exothermic in English. (Rachel, Year 8)

Rachel was also even more explicit than Louise in comparing *kanji* favourably to hiragana:

> If we know what the kanji is we get a better understanding of what the context is than doing hiragana because when hiragana doesn't actually give – kanji's like a character, it's like a picture, but whereas hiragana's just letters, so instead of saying the word cat, like C A T, we can say cat as in a picture, so we might get a better understanding of the context than just the word itself. (Rachel, Year 8)

In interview Murdoc, another student, chose to talk about the way *kanji* contains meaning in relation to his Humanities class rather than the science class, but his focus was on learning the actual *kanji* rather than learning the concept:

> Here with kanji you can see with some of the radicals, where some could be placed, like with the kanji ash that I learnt today, it's got fire in it.
> (Murdoc, Year 8)

Murdoc most likely knew what *ash* meant in English, and this contrasts with Rachel's comment about learning a concept in English for the first time. Concepts that are difficult for students to understand in English in humanities can also be accessed through *kanji* and re-translated back into English, however. For example, delta can come back into English as a visual for 'sandbanks-in-a-river' via the relevant *kanji* (see Turner, 2015).

Audrey, the non-Japanese-speaking science teacher co-teaching with Louise, also stated her understanding that the incorporation of Japanese was helping the students learn science by giving an example:

> The word for organism which I say is a living thing, if you break up the kanji it actually means 'life thing'. [That's] the actual meaning. [...] I am very much for [CLIL]. The students [...] not only understand the science but they're understanding it and remembering it in different ways. (Audrey, interview)

When asked whether she thought Japanese was slowing the students down in the science class, she answered:

> No, I don't. I think if anything it's actually helping them understand in a deeper way, the deeper meaning of what it is. (Audrey, interview)

Louise also considered this to be the case:

> [Students] can make more connections with the content itself. By accessing it through two languages. (Louise, interview)

Therefore, it appeared that, in the case of *kanji,* the inclusion of Japanese language was strongly content-driven, and Louise put some thought into how the use of Japanese in particular – rather than a foreign language in general – could help the students' understanding of content.

4.2   Learning Japanese and learning science

Although *kanji* was perceived as a useful tool for learning concepts and integrated into content learning, it appeared that it occupied an intermediary Japanese-English – rather than fully Japanese – space for participants. This related to the way the *kanji* helped to break down or reposition concepts when translated (back) into English. In discussions about learning and assessment in general, Japanese appeared to be conceptualized as something separate from the learning of science. The eight students who completed the survey all reported themselves to be satisfied with the amount of Japanese language use in class, and it was clear from observation that the students were engaged in the Japanese component of their science class. However, science appeared to be equated with English in the students' minds. The quotation below from Grace evidences this:

> In science, we were just learning about sexual reproduction, and I felt like it was more of a balance between Japanese and the subject itself. It's more fascinating to see that they've – our teachers have [...] balanced it out. So it's not just we'll stick to Japanese a hundred percent on one day. We would be able to balance it out so we'd actually learn from both the language and the subject itself.   (Grace, Year 8)

'The subject itself' in the excerpt above appeared to be the responsibility of Audrey, the non-Japanese speaking teacher:

> [The students] think I'm the science teacher, I do lead a lot more, I tend to be more with behavioural management, maybe they view me as that sort of image, if it comes from me, they think they've got to do this.
> (Audrey, 1st after-class reflection)

> [The content area] pressure's kind of taken off me, and I can just focus on how can I bring the language into this? You know, [...] what can we do in this class with this topic that would help the students to retain the language, to understand things more? (Louise, interview)

This separation between the core learning of science and Japanese was further evidenced by the students' discussion of assessment in their interviews. The main assessment was delivered in English with a supplementary page in Japanese. One student – Kim – reported the presence of Japanese on the exams to be negative:

> I think that the way we learn it is good, but I don't think that there should be as much Japanese on our [...] exams. [...] I think that for exams we should focus more on the subject because that's what exams are about, learning the subject.
> (Kim, Year 8)

The fact that Japanese could be considered to obscure subject area learning, underpins this quotation and, even though the other students professed themselves to be satisfied with the quantity of Japanese on the assessments, their quotations below also indicate the separation in their minds. In the final quotation, Grace was talking about Humanities and not science, but the quotation is included to show her way of thinking about Japanese and subject area assessment:

> [The Japanese part is] like about one page max. Probably double-sided or something, but [it's] all the Japanese that we've already been taught and revised over, so it's pretty easy for the Japanese part.
> (Rachel, Year 8)

> We have the normal – say we're doing a science test, we'd have the normal English test and then attached would be an extra Japanese sheet. So we could look at – we'd learn it – take the Japanese that we've learnt from the class and then answer on the test, and I like that way of being assessed.
> (Chad, Year 8)

> With our Humanities, our teacher [...], his assessments are more based on the subject itself, but I think it's also a good way because when we do our assessments it's based more on the subject, but when we're doing like class work and things he involves more Japanese.
> (Grace, Year 8)

Table 3. Excerpt from main science assessment on physical and chemical change

6   Describe the particles in a solid.
   A.   The particles in a solid are strongly bonded to each other so the solid has a definite shape.
   B.   The particles in a solid are strongly bonded to each other so the solid does not have a definite shape.
   C.   The particles in a solid are weakly bonded to each other so the solid has a definite shape.
   D.   The particles in a solid are weakly bonded to each other so the solid does not have a definite shape.

Louise also chose to discuss the potential language barrier when talking about assessment:

Section 7

Mark the following as TRUE （○） or FALSE （×） *(8 marks)*

1. 水は氷(こおり)になります。これは物理変化です。＿＿
2. ケーキを作(つく)ります。これは化学変化です。＿＿
3. 水と塩(しお)(sugar)を混(ま)ぜます(mix)。これは化学変化です。＿＿
4. 気体は液体になります。これは物理変化です。＿＿
5. 固体は気体になります。これは化学変化です。＿＿
6. 二つの液体を混ぜて、新(あたら)しい赤(あか)い固体が見えます。これは物理変化です。＿＿

Figure 3. Excerpt from additional assessment in Japanese on physical and chemical change

> If I am, say, assessing the content in the target language, like if I'm asking questions about reproduction in Japanese, and [the students] get some of those questions wrong on the test, can I assume that that's because they don't know the content? Or was that a language barrier? I know that I taught them those words, but it doesn't mean that they picked up those words. […] Should I test those words first [before putting them in context in the assessment]? I sort of have that slight worry of assessing fully in the language. (Louise, interview)

Louise was observed to give a warm-up quiz to students about physical and chemical change and, as she reported, it was clear that she was teaching and reinforcing important words and concepts in Japanese. However, her concern was understandable given the priority attributed to the subject area and the way that Japanese was positioned in a supporting role.

4.3 Explanation and application

Although Japanese language integration was found to be important in the class, ensuring students understood the science was considered most important and both the teachers and students appeared to believe that Japanese might interfere with explanations as well as assessments. *Kanji* was found to be beneficial, but if students did not understand explanations, English was found to become the language of choice, and Japanese was more firmly positioned as the language of application. Audrey in particular reported communicative language use as an obstacle:

> We have the common goal of teaching the language, the goal is to bring the two together […]. But like today, where it's really new things for them and it's a concept that they can't see, it's probably going to be more English […] for them to understand it, and then we'll use Japanese more when we're applying it. They understand in their native language best. (Audrey, 1st after-class reflection)

In order to help students to apply their knowledge using Japanese, Louise focused on teaching structures that were relevant to the science that was being taught. The extract from interview gives an example of this:

> I try to get [the students] to write basic sentences, like for example in their practicals they'll have to describe what happens. I think that describing is one of the easier things that they can begin to learn. So, teaching them the verbs and the structure that they might do for a simple sentence to describe, get them to write a few things.
> (Louise, interview)

A relevant illustration of the students' work in Japanese is provided below. In this activity, students made a video of themselves describing a physical and chemical change in Japanese. The transcript is a written version of what students said on a video. Although a written text was not sighted, Kim was reading in the activity:

> Kim: Ice cream sandwich
> アイスクリームサンドイッチを作ります。バニラとeggとflourとアイスクリームを混ぜます。次に私はトレーにクッキーを置きます。Ovenで焼きます。アイスクリームは物理変化です。クッキーは科学変化です。美味しい。
> [I will make an ice cream sandwich. I will mix vanilla, egg, flour and ice-cream. Next, I will put cookies on a tray. I will cook it in the oven. The ice-cream will undergo a physical change. The cookies will undergo a chemical change. Yum.]
> Students: Cute! [Kim], so cute!

Application also extended to classroom expressions in Japanese. Louise was using, and encouraging students' use of, functional classroom language, and Audrey joined in, understanding this to be good modelling for the students:

> I'm like the actual students in the room, and [...] I'm kind of a good model I guess for them, because [...] I'm showing that I'm trying as well, and just sometimes I'll surprise them and I'll say [...] you know like 'stand up' [...]. I'll greet the classroom or I'll dismiss the class in Japanese and they enjoy that, and they giggle cause it's funny me saying it, but I think they appreciate the fact that [...] I'm kind of, you know, trying to learn as well.
> (Audrey, interview)

Thus, although communication in English was prioritised if students had trouble with the science, Japanese language use appeared to be attributed particular roles, such as the application of learning and routine instructions.

5. Discussion and conclusion

This article has focused on two significant contextual issues in the implementation of CLIL in a science classroom: (1) the planned use of more than one language (Japanese and English in this case) rather than the sustained instructional use of only the target language, and (2) the nature of the target language. Without external guidelines around the extent to which the target language is to be the language of instruction, teachers need to make decisions about how much to incorporate the language into class. In the study, the Japanese language teacher was found to experiment with the use of Japanese in ways that helped students to engage with scientific content and simple language forms.

In the science classroom, the teacher used Japanese in both explanation and application. For explanation in particular, she took advantage of a linguistic benefit of *kanji*: the characters could be translated back into simpler – or just different – English to help explain scientific concepts to the students. The fact that English was the first language of the teacher may have helped her to see and prioritise this benefit because she herself was able to understand concepts she had learned in English more deeply when she read the *kanji*. She was able to address two of Jäppinen's (2005, p. 153–154) five critical discovery learning areas through this approach: 'awareness of existing concepts and the ability to understand, use, apply, and explain them' and 'awareness of meaning schemes and the ability to create links between them'. The way the non-Japanese speaking science teacher co-teaching with the Japanese language teacher viewed this aspect of *kanji* as beneficial to the students' deeper understanding of scientific concepts further indicated the relevance of Japanese to the science classroom that went beyond the learning of Japanese *per se*. This relevance extends to other character-based languages, such as Chinese.

Nevertheless, there seemed to be a separation in the students' – and non-Japanese speaking science teacher's – minds between more communicative Japanese language use and learning science. Japanese was considered to be important but learning science, and displaying this knowledge via assessments, was still considered to belong more firmly to the English domain. Japanese was observed to be used in warm-ups and during the application phase of learning, and the Japanese language teacher taught language structures that would suit application of what students were learning. Literacy was used as a further way to increase students' oral confidence by giving them something to read out loud. Students were assessed on the Japanese component of their learning in science on an extra sheet attached to the general assessment, and informal feedback was given on their progress in Japanese language. The separation of Japanese language use and learning science may have been connected to students' very limited communicative proficiency in Japanese and the way that Japanese language use, for them, could generally be equated with 'learning Japanese'. Although, as Nikula, Dalton-Puffer, Llinares, & Lorenzo (2016, p. 10) pointed out, 'CLIL and other forms of bilingual education involving integrated language and content teaching challenge the often taken for granted separation between "language subjects" and "content subjects"', this separation can still, to some degree, be evidenced in CLIL classrooms. In the study, when integration was clear, such as with the use of *kanji*, language was being harnessed as a way to learn science, but for conceptual understanding rather than to practice and extend discipline-based literacies (see Meyer et al., 2015).

Finally, the use of *kanji* can be linked to the *kokuji mondai*, or the complexity of needing to learn three scripts in Japanese. The Japanese language teacher appeared to allow both content and her students' limited proficiency (and limited opportunities for exposure to Japanese in everyday life) to guide her in addressing the way she approached literacy and oracy. The content provided an opportunity to show the benefits of *kanji* as a writing system, and *kanji* recognition, rather than production, was the main focus. This can be compared with Lee and Lee's (2017) recognition of the significance of 'idiosyncratic features' of the language of instruction. The teacher then used *hiragana* to assist students in framing their thinking and learning, with one end point being oracy. Given the low proficiency level of the students, the use of language was generally very controlled, but the approach taken by the teacher demonstrated that she was experimenting and seizing opportunities in a way that both the students and her co-teacher appreciated. Given the variation between the linguistic and cultural characteristics of different languages and the way the learning of content – and language itself – can subsequently be stimulated in different ways, approaches that harness language-specific features, such as the use of *kanji*, are worthy of further investigation. The study further suggests the need to guide teachers' navigation of an integrated curriculum in settings where students have only a very rudimentary understanding of the target language and a trajectory that requires them to learn the CLIL subject areas solely through the medium of the dominant language (English) in senior secondary school.

References

Abe, Y. (2002). 漢字という障害 [An Obstacle Named Kanji]. *Shakai gengogaku*, 2, 37–55.
Abe, Y. (2015). ことばのバリアフリー: 情報保障とコミュニケーションの障害学 [Barrier-free Language: Disability studies of information assurance and communication]. Tokyo: Seikatsu shoin.

Andō, M. (1942). 日本語のむづかしさ [The difficulty of Japanese language]. *Nihongo*, 2(3), 4–11.

Braun, V., & Clarke, V. (2006). Using thematic analysis in psychology. *Qualitative Research in Psychology*, 3(2), 77–101. https://doi.org/10.1191/1478088706qp063oa

Bravo, M.A. (2017). Cultivating teacher knowledge of the role of language in science: A model of elementary grade pre-service teacher preparation. In A.W. Oliveira & M.H. Weinburgh (Eds.), *Science teacher preparation in content-based second language acquisition* (pp. 25–40). Springer: Switzerland. https://doi.org/10.1007/978-3-319-43516-9_2

Cenoz, J., Genesee, F., & Gorter, D. (2014). Critical analysis of CLIL: Taking stock and looking forward. *Applied Linguistics*, 35(3), 243–262. https://doi.org/10.1093/applin/amt011

Coyle, D. (2007). Content and language integrated learning: Towards a connected research agenda for CLIL pedagogies. *International Journal of Bilingual Education and Bilingualism*, 10(5), 543–562. https://doi.org/10.2167/beb459.0

Coyle, D., Hood, P., & Marsh, D. (2010). *CLIL: Content and language integrated learning*. Cambridge: Cambridge University Press.

Cross, R., & Gearon, M. (2013). *Research and evaluation of the content and language integrated learning (CLIL) approach to teaching and learning languages in Victorian schools*. Melbourne Australia: Melbourne Graduate School of Education, The University of Melbourne.

Dalton-Puffer, C. (2013). A construct of cognitive discourse functions for conceptualizing content and language integration in CLIL and multilingual education. *European Journal of Applied Linguistics*, 1(2), 216–253. https://doi.org/10.1515/eujal-2013-0011

Dalton-Puffer, C., Llinares, A., Lorenzo, F., & Nikula, T. (2014). 'You can stand under my umbrella': Immersion, CLIL and bilingual education. A response to Cenoz, Genesee & Gorter (2013). *Applied Linguistics*, 35(2), 213–218. https://doi.org/10.1093/applin/amu010

Dalton-Puffer, C. (2016). Cognitive discourse functions: Specifying an integrative interdisciplinary construct. In E. Dafouz & T. Nikula (Eds.), *Conceptualising integration in CLIL and multilingual education* (pp. 29–54). Bristol: Multilingual Matters. https://doi.org/10.21832/9781783096145-005

De Jong, E. (2002). Effective bilingual education: From theory to academic achievement in a two-way bilingual program. *Bilingual Research Journal*, 26(1), 65–84. https://doi.org/10.1080/15235882.2002.10668699

De Kretser, A., & Spence-Brown, R. (2010). *The current state of Japanese language education in Australian schools*. Melbourne: Education Services Australia.

Department of Education and Training (DET). (2017). *Languages provision in Victorian Government schools*, 2016. Retrieved from <http://www.education.vic.gov.au/Documents/school/teachers/teachingresources/discipline/languages/2016_Languages_provision_report.pdf>

Eurydice. (2006). *Content and Language Integrated Learning (CLIL) at School in Europe*. Eurydice. Retrieved from http://www.indire.it/lucabas/lkmw_file/eurydice/CLIL_EN.pdf

Fernández-Sanjurjo, J., Fernández-Costales, A., & Arias Blanco, J.M. (2017). Analysing students' content-learning in science in CLIL vs. non-CLIL programmes: Empirical evidence from Spain. *International Journal of Bilingual Education and Bilingualism*, 1–14. https://doi.org/10.1080/13670050.2017.1294142

Fielding, R., & Harbon, L. (2015). Implementing a content and language integrated learning program in New South Wales primary schools: Teachers' perceptions of the challenges and opportunities. *Babel*, 49(2), 16+

Fortune, T.W., & Tedick, D.J. (2008). One-way, two-way and indigenous immersion: A call for cross-fertilization. In T.W. Fortune & D.J. Tedick (Eds.), *Pathways to multilingualism: Evolving perspectives on immersion education* (pp. 3–21). Clevedon: Multilingual Matters. https://doi.org/10.21832/9781847690371-004

Genesee, F. (1987). *Learning through two languages: Studies of immersion and bilingual children.* Cambridge, MA: Newbury House.

Genesee, F. (2004). *Dual language development and disorders: A handbook on bilingualism and second language learning.* Baltimore, MD: Paul H. Brooks.

Halliday, M.A.K., & Mattheissen, C. (2004). *An introduction to functional grammar* (3rd edition), London: Arnold.

Hovhannisyan, A. (2018). Japanese language education in the greater East Asia co-prosperity sphere and the Kokuji Mondai (National Script Problem). In K. Hashimoto (Ed.), *Japanese Language and Soft Power in Asia* (pp. 65–81). Singapore: Palgrave MacMillan. https://doi.org/10.1007/978-981-10-5086-2_4

Hüttner, J., Dalton-Puffer, C., & Smit, U. (2013). The power of beliefs: Lay theories and their influence on the implementation of CLIL programmes. *International Journal of Bilingual Education and Bilingualism*, 16(3), 267–284. https://doi.org/10.1080/13670050.2013.777385

Jäppinen, A. (2005). Thinking and content learning of mathematics and science as cognitional development in content and language integrated learning (CLIL): Teaching through a foreign language in Finland. *Language and Education*, 19(2), 147–168. https://doi.org/10.1080/09500780508668671

Lee, W., & Lee, J.S. (2017). Math instruction is not universal: Language specific pedagogical knowledge in Korean/English two-way immersion programs. *Bilingual Research Journal*. https://doi.org/10.1080/15235882.2017.1380729

Lindholm-Leary, K. (2001). *Dual language education.* Clevedon, UK: Multilingual Matters. https://doi.org/10.21832/9781853595332

Lindholm-Leary, K., & Genesee, F. (2010). Alternative educational programs for English language learners. In California Department of Education (Eds.), *Improving education for English learners: Research-based approaches* (pp. 323–382). Sacramento, CA: CDE Press.

Lyster, R. (2007). *Learning and teaching languages through content: A counterbalanced approach.* Amsterdam: John Benjamins. https://doi.org/10.1075/lllt.18

Lyster, R., & Mori, H. (2006). Interactional feedback and instructional counterbalance. *Studies in Second Language Acquisition*, 28(2), 321–341. https://doi.org/10.1017/S0272263106060128

Martin-Beltrán, M. (2010). The two-way language bridge: Co-constructing bilingual language learning opportunities. *Modern Language Journal*, 94(2), 254–277. https://doi.org/10.1111/j.1540-4781.2010.01020.x

Meyer, O., Coyle, D., Halbach, A., Schuck, K., & Ting, T. (2015). A pluriliteracies approach to content and language integrated learning – Mapping learner progressions in knowledge construction and meaning-making. *Language, Culture and Curriculum*, 28(1), 41–57. https://doi.org/10.1080/07908318.2014.1000924

Nikula, T. (2015). Hands-on tasks in CLIL science classrooms as sites for subject-specific language use and learning. *System*, 54, 14–27. https://doi.org/10.1016/j.system.2015.04.003

Nikula, T., Dalton-Puffer, C., Llinares, A., & Lorenzo, F. (2016). More than content and language: The complexity of integration in CLIL and bilingual education. In T. Nikula, C. Dalton-Puffer, A. Llinares, & F. Lorenzo (Eds.), *Conceptualising integration in CLIL and multilingual education* (pp. 1–25). Clevedon, UK: Multilingual Matters. https://doi.org/10.21832/9781783096145-004

Oller, D. K., & Eilers, R. (2002). *Language and literacy in bilingual children.* Clevedon, UK: Multilingual Matters. https://doi.org/10.21832/9781853595721

Paran, A. (2013). Content and language integrated learning: Panacea or policy borrowing myth? *Applied Linguistics Review*, 4(2), 317–342. https://doi.org/10.1515/applirev-2013-0014

Pérez-Cañado, M. L. (2012). CLIL research in Europe: Past, present, and future. *International Journal of Bilingual Education and Bilingualism*, 15(3), 315–341. https://doi.org/10.1080/13670050.2011.630064

Pérez-Vidal, C., & Roquet, H. (2015). The linguistic impact of a CLIL science programme: An analysis measuring relative gains. *System*, 54, 80–90. https://doi.org/10.1016/j.system.2015.05.004

Rasulo, M., de Meo, A., & de Santo, M. (2017). Processing science through content and language integrated learning (CLIL): A teacher's practicum. In A. W. Oliveira & M. H. Weinburgh (Eds.), *Science teacher preparation in content-based second language acquisition* (pp. 305–322). Springer: Switzerland. https://doi.org/10.1007/978-3-319-43516-9_17

Oliveira, L. C. (2017). A language-based approach to content instruction (LACI) in science for English language learners. In A. W. Oliveira & M. H. Weinburgh (Eds.), *Science teacher preparation in content-based second language acquisition* (pp. 41–56). Springer: Switzerland. https://doi.org/10.1007/978-3-319-43516-9_3

Silva, C., Weinburgh, M., Malloy, R., Horak Smith, K., & Nettles Marshall, J. (2012). Toward integration: An instructional model of science and academic language. *Childhood Education*, March/April, 91–95. https://doi.org/10.1080/00094056.2012.662119

Smala, S. (2016). CLIL in Queensland: The evolution of 'immersion'. *Babel*, 50(2–3), 20.

Swain, M. (1985). Communicative competence: Some roles of comprehensible input and comprehensible output in its development. In S. Gass & C. Madden (Eds.), *Input in second language acquisition* (pp. 235–253). Rowley, MA: Newbury House.

Swain, M. (2000). French immersion research in Canada: Recent contributions to SLA and applied linguistics. *Annual Review of Applied Linguistics*, 20, 199–212. https://doi.org/10.1017/S0267190500200123

Tanaka, K. (2011). 漢字が日本語をほろぼす [Kanji are destroying Japanese language]. Tokyo: Kadokawa SSC Shinsho.

Tedick, D. J., & Young, A. I. (2014). Fifth grade two-way immersion students' responses to form-focused instruction. *Applied Linguistics*, 37(6), 784–807. https://doi.org/10.1093/applin/amu066

Thomas, W. P., & Collier, V. P. (2002). *A national study of school effectiveness for language minority students' long-term academic achievement: Final report.* Santa Cruz, CA/Washington, DC: Center for Research on Education, Diversity & Excellence.

Turner, M. (2013). Content-based Japanese language teaching in Australian schools: Is CLIL a good fit? *Japanese Studies*, 33(3), 315–330. https://doi.org/10.1080/10371397.2013.846211

Turner, M. (2015). The significance of affordances on teachers' choices: Embedding Japanese across the curriculum in Australian secondary schools. *Language, Culture and Curriculum*, 28(3), 276–290. https://doi.org/10.1080/07908318.2015.1085063

Ullmann, M. (1999). History and geography through French: CLIL in a UK secondary school. In J. Masih (Ed.), *Learning through a foreign language: Models, methods and outcomes* (pp. 96–105). London: Centre for Information on Language Teaching and Research.

Weinburgh, M.H., & Silva, C. (2010). *Science content knowledge and language acquisition: Replacing, reloading, repositioning, revealing and retiring academic words.* Paper presented at the annual meeting of the Association for Science Teacher Education, Sacramento, CA.

Wode, H. (1999). Language learning in European immersion classes. In J. Masih (Ed.), *Learning through a foreign language: Models, methods and outcomes* (pp. 16–25). London: Centre for Information on Language Teaching and Research.

Teacher language awareness and scaffolded interaction in CLIL science classrooms

Daozhi Xu[1,2] and Gary James Harfitt[3]
[1] Sun Yat-sen University | [2] University of Tasmania |
[3] The University of Hong Kong

Teacher language awareness (TLA) constitutes the teacher's self-reflective knowledge about the operation of language systems in pedagogical practices. This study focuses on teachers' understanding of learning of language and learning through language in Content and Language Integrated Learning (CLIL) contexts, exploring how teachers proceduralise their knowledge of language to facilitate science learning in Hong Kong. By analysing the reflective relationship between TLA and scaffolding strategies of two teachers (students $n = 31$; 32) during a set of lessons in a secondary school, this paper suggests that it is critical to re-orient the TLA focus from teachers to the act of learning and learners' needs. This expanded conceptual framework of TLA sheds light on how to transform teachers' implicit knowledge of language into explicit awareness of scaffolding in class. The TLA-filtered, scaffolded interactions can therefore promote the use of language not merely for pedagogical purposes but also as a cognitive learning tool.

Keywords: Teacher language awareness, Content and Language Integrated Learning (CLIL), scaffolding strategies, learners' needs, science classrooms

1. Introduction

Content and Language Integrated Learning (CLIL) education aims to achieve competence in both content subject and English language knowledge. Emerging in the 1990s, this "dual focused educational approach in which an additional language is used for the learning and teaching of both content and language" has a worldwide application (Coyle, Hood, & Marsh, 2010, p. 1; Merino & Lasagabaster, 2018). CLIL has gained increasing prominence particularly in Europe and Asia due to its popular double objectives and design (Graham, Choi, Davoodi, Razmeh, & Dixon, 2018). It can also be a product of local language policies, for example in Hong Kong which provides a case study for this research.

For a number of years before Hong Kong's return to Chinese sovereignty in 1997, a laissez-faire approach had been adopted regarding the medium of instruction (MOI) in each school. While some secondary schools adopted the students' mother tongue, namely Chinese (Cantonese), as the Medium of Instruction (CMI), the vast majority claimed to be English-medium (EMI). In reality, however, most of the so-called EMI secondary schools employed an ad hoc mixture of Chinese and English ("mixed code") in the teaching of content subjects. While the Government had encouraged teaching through the students' mother tongue, in 1997/98 the Government issued *Medium of Instruction Guidance for Secondary Schools* (which was referred to as "Firm" Guidance): only 114 public-sector secondary schools, amounting to 30% of Hong Kong's total 421 secondary schools, would be EMI (and then only if they fulfilled certain conditions relating to their student intake and their staff); the remaining 70% would be CMI and it was compulsory for them to use CMI at the junior secondary level (Evans, 2013). This well-intentioned attempt to provide clarity and promote mother-tongue teaching proved to be unpopular in many quarters: for instance, among aspirational parents, many of whom who perceived English as having more social and cultural value than Chinese, and among principals and communities of CMI schools, who felt that, given the socio-economic importance of English in the Hong Kong context, being identified as a CMI school was like being labelled "second-class" (Chan, 2016; Choi, 2003). In 2010/11, against this background, the Government started to implement a fine-tuning MOI policy (Education Bureau, 2009). Under this policy, schools would no longer be labelled CMI. Those schools that were previously designated CMI were given greater autonomy to determine the MOI for content subjects. One of the Government's key objectives was to provide students in former CMI schools with greater exposure to English. As a result of this change, there has been a marked increase in the teaching of content subjects (especially Science and Mathematics) through English, though the teaching modes vary considerably, i.e. in practice, the same subject can be conducted primarily in Chinese and sometimes in English for certain units of the subject in some schools whereas in some others, the subject can be taught entirely in English (or in Chinese). The fine-tuned MOI policy may have been politically expedient, allaying the concerns of many parents and principals, but it has also created challenges for the teachers of content subjects (many of whom had little or no previous experience of teaching through EMI) and for their students (Fung & Yip, 2014).

In CLIL programs, language plays a crucial role as both the medium and the object of instruction, which calls for language-aware content teachers (Andrews & Lin, 2017; Liyanage & Bartlett, 2010). Such a need is obvious in Hong Kong, not only because expressing and delivering content knowledge in the classroom is realised through language during the talk-in-interaction between the teacher and students, but also because most CLIL teachers are non-native speakers of English and their command of English language varies (Lasagabaster & Sierra, 2016). It is not the case that native English-speaking teachers (NEST) would definitely outperform non-native English-speaking teachers (NNEST) in CLIL classrooms, because the bilingual skills of NNEST might enable them to perceive the difference between languages and perhaps teach more effectively the content subject in L2. Nevertheless, the question remains as to what levels of language proficiency is required for CLIL teachers to facilitate meaning making in content subject classrooms (Morton, 2018), particularly for those who used to teach their subjects in Chinese but now need to shift to English after the implementation of the fine-tuned MOI policy. Furthermore, while CLIL teachers are expected to enhance students' English competency which constitutes one of the key teaching goals, many teachers identify themselves as content subject teachers only (Koopman, Skeet, & de Graaff, 2014; Lo, 2019; Tan, 2011). A considerable proportion of these teachers are seen to have a weak grasp of language teaching knowledge and professional skills to develop students' English competency (Hoare, 2003). Despite efforts have been made, the language focus in the CLIL classrooms is largely confined to rote learning of technical terms and drilling of language-related exercises (e.g., blank filling exercises and low-level comprehension exercises) in a piecemeal fashion (Lyster, 2007; Regalla, 2012).

2. Teacher language awareness

Both "learning language" and "learning through language" are important for the cognitive development of children (Halliday, 1993, p. 93). These two interrelated aspects underline the double objectives of CLIL education. Addressing the significant role of language in the learning/teaching process, the framework of Teacher Language Awareness (TLA) sheds light on "the interface between what teachers know, or need to know, about language and their pedagogical practice" (Andrews & Svalberg, 2017, p. 220). TLA is believed to affect the effective language input and output in classes (Andrews, 2007). As "an essential attribute of any competent L2 teacher," TLA has received considerable attention in the studies of English-as-Second/Foreign-Language (ESL/EFL) classrooms, providing insight into teacher cognition and knowledge about the operation of language systems in the pedagogical process (Andrews, 2007, p. ix; Thornbury, 1997). TLA is potentially significant in influencing how language is instructed, organised, and learnt in the ESL/EFL classrooms. Studies show that it is equally relevant to CLIL teachers (Andrews & Lin, 2018; García, 2009; Lindahl, Baecher, & Tomas, 2013). As such, however, there remains a paucity of empirical study in CLIL contexts on how to enhance the TLA of content subject teachers and thereby improve their pedagogical practices.

According to Andrews, TLA involves three aspects: first, a teacher's L2 language proficiency; second, a teacher's mastery of the specialised language of subject matter; and third, the teacher's knowledge of learners' language proficiency, especially related to subject matter (Andrews, 2007; Andrews & Lin, 2018). Linking TLA to the Pedagogical Content Knowledge (PCK) which is an "amalgam of content and pedagogy" (Shulman, 1987, p. 8), Andrews (2007) suggests that TLA forms "a pedagogically related reflective dimension of language proficiency" and "a sub-component of the L2 teacher's PCK, which interacts with the other sub-components" (p. 30). In this respect, Andrews (2007) suggests that TLA consists of "declarative and procedural dimensions;" the former designates the teacher's "possession of [language-related] knowledge" and the latter is linked to the pedagogical strategies regarding "the use made of such knowledge" in class (p. 31). Both dimensions are interrelated, but there remains a challenge of realising the transference of TLA from teachers' knowledge of language (*declarative* dimension) to the use of knowledge (*procedural* dimension) in classroom interaction (Andrews & Lin, 2018).

It is vital that teachers transform their implicit understanding of language-related knowledge into an explicit awareness that informs pedagogical practices. Building upon the framework proposed by Edge (1988) which highlights L2 teachers' three roles as language *user*, language *analyst*, and language *teacher*, three domains of TLA were further developed in Wright and Bolitho (1993) and in Lindahl (2013, 2016), to address the need of teachers to develop language knowledge and procedural awareness. The three domains are (a) the *user* domain, which refers to the teacher's English language proficiency and his/her use of English to communicate with people who speak other dialects of English or other languages; (b) the *analyst* domain, which consists of the teacher's understanding of grammars and rules by which language works in the subject-specific, academic contexts; and (c) the *teacher* domain, which refers to the pedagogical knowledge of language teaching (Andrews & Lin, 2018; Lindahl & Watkins, 2015). This theory sheds light on how TLA can be enhanced as the teacher shifts across different identities as users, analysts and teachers of language. But a salient question raised by Andrews and Lin (2018) highlights the connection between TLA and learning: "How might a language-aware teacher be better equipped to enact the curriculum in ways that support student learning?" (p. 60). While the three domains of TLA place a spotlight on how teachers consider their treatment of language, more attention need to be paid to their understanding of learners, the learning process, and contextual factors related to the language in class, which are critical to guide the teacher's language teaching (Andrews & Lin, 2018; Lindahl & Watkins, 2015). Teaching and learning are always integrated. The interface between teaching and learning underlines the transformation between the teacher's declarative knowledge and the corresponding procedural knowledge of language. In other words, learning of language as understood by teachers can direct them to provide effective, targeted scaffolding for students in the classroom learning process. It is worthwhile to examine how teachers understand the use of language from the perspectives of learners and learning (Andrews & Lin, 2018). The enactment of TLA can thereby facilitate students' learning of language in CLIL programs.

It is acknowledged that TLA influences various aspects of pedagogical practices (Andrews, 2007). In a sense, TLA affects nearly "every decision that the L2 teacher makes in relation to the language made available for learning" (Andrews & Lin, 2018, p. 60). TLA serves as a "filter" which guides and monitors the classroom talk, identifies learning gaps, and intervenes with appropriate scaffolding and mediation strategies (Andrews, 2007, p. 38). While the impact of TLA upon pedagogical practices is never straightforward because multiple factors including teachers' preparation, classroom organisations and students' responses would intervene in the operation of TLA (Andrews, 2007), it is meaningful to elucidate the intersections between TLA and scaffolded interactions. Given that cognition and awareness can be elusive and difficult to judge and assess, a mapped-out connection between the procedural TLA in cognition and scaffolding strategies in practice provides insight into teacher professional development.

To this end, this study focuses on teachers' understanding of learning of language and through language in CLIL science classrooms, with a purpose of exploring how teachers proceduralise their knowledge of language to facilitate learning. Here, learning of language refers to the English language (L2) learning in CLIL contexts. Learning through language means the learning of content-related, specialised language with an aim of expressing scientific ideas, reasoning and solving problems. Both aspects of language learning – the L2 and the language of content subject – are integral in CLIL programs. The study reported in this paper provides implications for teachers' pre-service and in-service training of language awareness. It will shed light on how TLA informs pedagogical practices and is conducive to the learning of and through language in CLIL contexts.

3. Methodology

This study sits within a wider research context and project investigating the role of English language in EMI content subject lessons in Hong Kong. The primary research questions are:

1. What is teachers' awareness of language-related issues in the CLIL learning of science?
2. What strategies of scaffolding are employed by teachers to facilitate students' language learning in the classrooms?

This paper reports the findings of teacher language awareness and scaffolding strategies of two teachers in a set of consecutive lessons to conduct a mini-science project "Building a Frame Structure" among two classes of Form 1 (Grade 7) students. In the observed science project, two classes of students were guided to design and build a frame structure with drinking straws and adhesive tapes. In each class, students were divided into 8 groups (3–4 pupils in each group) and the group that built the frame structure which could hold the largest amount of weight per gram of its own weight was deemed the winner of the competition. Table 1 presents the observed lessons and two classes participating in this mini-science project. The purpose of choosing a set of project-based lessons for this study is out of the consideration to maintain consistency and coherence of teaching contents, useful for the holistic analysis of classroom interactions (see Gibbons, 2003, p. 255).

Table 1. Class A and B

Mini science project	Class A (TA)	Class B (TB)
Number of lessons (45 minutes per lesson)	6	8
Number of students	31	32

3.1 Participants

Both teachers are from one school, classified as Band 1 (representing the top tier in the three-tier system of Hong Kong secondary schools). Table 2 details the two participant teachers' profiles.

Table 2. Teacher A and B

	Teacher A (TA)	Teacher B (TB)
Qualifications	Bachelor of Science – Chemistry & The Postgraduate Diploma in Education (PGDE)	Bachelor of Science – Animal and Plant Biotechnology
Years of experience	10 Years	8 Years
Medium of Instruction	EMI	EMI

Students from both classes are not streamed, which means that they were not distributed to classes because of their academic ability. When data was collected, these students had only spent one semester in the secondary school. Researchers did not observe considerable difference in the academic abilities across the two classes. Teachers and students participating in this study are ethnically Chinese, which reflects the make-up of local Hong Kong classes. As this study is aimed at examining teachers' language awareness and its interrelations with scaffolding practices, it is not our intention to compare two teachers or two classes, but to have more than one teacher/class enlarges our database and presents a more balanced view of the CLIL situations in Hong Kong.

To understand the learners' perspectives of language learning, which provides a basis for researchers to examine teachers' knowledge of learners, a survey of two classes of students ($n = 31; 32$) was conducted through a questionnaire. Seven questions in the student questionnaire cover different aspects related to their language learning, including their mother tongue, language(s) spoken at home, language(s) used in primary schools, the difficulty of learning science in English, and their language preference when studying science. As the result of questionnaire shows, Cantonese is the mother tongue and home language of the majority students in both classes. In terms of the language used in the lessons of General Studies (the subject content of which involves science) in their primary schools, Cantonese rates above 90% and only three students in two classes had English as MOI in General Studies. For the question, "At this stage, do you have any difficulty learning science in English?", 23.3% students in Class A and 22.6% students in Class B respectively report that they have difficulty in general. As for the question which asks students to identify which aspect(s) of science learning in English is deemed most difficult (multiple choices), the highest response rate was awarded to the item "understanding scientific terms and/or concepts in English" which amounts to 78.6% and 76.9% of students in Class A and B, respectively. For students in Class A, over 30% of students find it difficult to ask and to answer questions in English in science lessons, and to engage discussions with classmates in English. For students in Class B, 42.3% students believe it difficult to understand science teachers' instructions in English in class. Answering questions in English is also deemed difficult for nearly 50% of students in Class B. When being asked "if you have a choice, which language would you prefer to use when learning science in secondary school", 73.3% students in Class A choose "mainly English with some Cantonese" and this choice is also the most popular among students (45.2%) in Class B. Using purely English only amounts to 10% and 19.4% in Class A and B, respectively. Over one third of students in Class B would like to use "mainly Cantonese with some English" in learning science.

The similarity of statistics from these two classes reveal that most of the observed students have encountered various difficulties in learning science in English as they adapt to an EMI learning environment in the secondary school. Students' learning needs and difficulties related to the use of language call for language-aware content teachers who not only have sufficient content subject knowledge and English proficiency but also understand the language learning from learners' perspectives and the nature of language learning in CLIL contexts.

3.2 Procedures

Semi-structured interviews with two teachers centre on the teachers' views on the role of language in CLIL learning, the language learning among their students, and the language policy of the school and the Education Bureau. Since TLA is arguably within the cognitive realm, to avoid teachers modifying their thoughts and awareness for the sake of answering the questions, the term TLA was not given to the two teachers during the interviews. The interviews were conducted in Cantonese (the teachers' mother tongue) and an interview protocol was followed to ensure consistency across two teachers and researchers. The excerpted interview data in this paper were translated by the researchers and then member checked by the two teachers so as to ensure the translated texts conveyed their original meaning. The lessons were video-recorded and transcribed verbatim for analysis.

3.3 Analysis

This study adopts a grounded approach to code teachers' awareness of language use in interview transcripts and scaffolding strategies emerging from the lessons. Coding was done by a research team of 5 individuals who conducted an iterative process of data checking across cases. Agreement was across the team as each member was familiar with the separate contexts and then read by the PIs and consultant advisor in her role as Chair Professor. The teachers' interview transcripts were first coded according to the relevance to the awareness of language-related issues. Then as the themes emerged which were related to learning of and through language and to the relations between learners and language, the codes were refined. The interview transcripts were then re-examined after the codes were finalised.

To facilitate the analysis of scaffolding in the classroom discourse, Sinclair and Coulthard's IRF model (1975), which comprises teacher initiation (I), pupil response (R), and teacher feedback or follow-up (F), was used to examine the interactive patterns between teacher and students. The codes of scaffolding strategies were drawn from the model of mediation by Gibbons (2003) between oral, everyday language and academic, specialist languages: (1) recasting to shift modes, (2) enabling students to reformulate, (3) requesting clarification, and (4) evoking personal knowledge, together with other recurring, salient themes identified in the co-construction of knowledge in the observed lessons. Besides, this paper draws on Holton and Clarke's construct of scaffolding (2006), which is categorised into different agents: expert scaffolding (teachers), reciprocal scaffolding (peers) and self-scaffolding (learners themselves), and two domains: conceptual and heuristic scaffoldings. An act of scaffolding often synthesises both aspects to facilitate problem-solving processes. It should be noted that this paper does not discuss the heuristic scaffoldings provided to solve the technical aspects of building a frame structure. It focuses on the language use, primarily via conceptual scaffoldings, to achieve co-construction of knowledge. When the codes were refined, the researchers went through the lessons transcripts to ensure the consistency and accuracy of codes.

4. Findings

With a focus on the teacher's understanding of learning of language and through language in CLIL science classrooms, the findings consist of the teachers' awareness of language from the perspectives of learners and learning, and their scaffolding practices in the classroom talk-in-interaction. The following sections include data and verbatim quotes from teachers' interviews and classroom transcripts.

4.1 L2 science teachers' awareness of language from the perspectives of learners and learning in CLIL contexts

The findings on teachers' language awareness in CLIL contexts include three interrelated aspects: the teacher's understanding of learning through language, of students' L2 language proficiency and language of subject matter, and of the impact of language policies on students' learning motivation and outcomes.

4.1.1 *Teachers' understanding of learning content subjects through language*

According to social constructivist theories of learning, there is a strong relationship between social interaction via the use of semiotic tools and individual acquisition of knowledge, skills and values (Vygotsky, 1978). Learning takes place in a social, dialogic environment of education where learners of different backgrounds engage in meaningful and collaborative activities. It is important to recognise the intricate relations between language and mental cognition. Language is instrumental in the cognitive construction of knowledge and experience, not only because one disseminates knowledge by using a language, but also because language forms the way in which one construes an experience and internalises it as knowledge. As Halliday argues, "language is the essential condition of knowing, the process by which experience becomes knowledge" (1993, p. 94). Reflecting on the central role of language in the learning process epitomises the meta-linguistic nature of TLA (Andrews, 2007, p. 29), concerning how language is used by students to express scientific ideas, to engage in tasks and activities, and to make meanings related to the subject of science. The meta-cognitive awareness reflects teacher's thinking of student-oriented approach in planning and implementing classroom talk. This is demonstrated by Teacher A's comments on the importance of group discussions in enhancing students' scientific understanding:

1. *(TA) Students have different views. Group discussions give them a chance to express their ideas, compare one's own view with others, critique each other's views so as to reach a most acceptable solution, etc. Discussions are a self-reflective thinking process for students, which may help them enhance their understanding of scientific knowledge.*

4.1.2 Teachers' knowledge about learners' English language proficiency and their cognitive knowledge of subject matter

This involves the teacher's understanding of (i) the curriculum required levels and (ii) the actual levels of (a) students' L2 language proficiency and (b) their cognitive knowledge of subject matter. The gaps between the required and the actual levels reveal the need for teachers' scaffolding. The teacher's awareness in this regard helps to identify at what stage and in what way s/he assists learners from diverse academic and language backgrounds to master the content subject knowledge and related English language expressions.

In this regard, Teacher B is keenly aware of the subtle differences between scientific terms, because when such terms appear the same in Chinese, confusion may arise among students. The accurate use of scientific terms matters in student's learning of science, while knowing "when" students can grasp different degrees of meaning at different learning stages is also important. In the interview, he gives examples of the progressive approach to teaching "weight" and "mass" (see quote 2). He also relates this awareness of subject matter knowledge, such as the difference between "secrete" and "release", to his professional training of biology (see quote 3). It is evident that students' knowledge, in terms of their conceptual understanding of science and the possible intervention of their mother tongue, informs the teacher when and how to impart his scientific knowledge to students.

2. (TB) *I tell Form 1 students to "measure the weight". Technically, it is not the "weight", but "mass". But students may not understand this during their first year of secondary school. It is good for them to have a general idea first; in their second year, I will then elaborate on the concept of "mass" and then let them know that the term "mass" is scientifically correct in this context. I think this [progressive] approach is easier for students to understand scientific concepts.*
3. (TB) *I am aware of the use of subject and verbs. Biology is my major for Bachelor's degree and we do need to pay attention to the subtle difference in the meaning of scientific terms. One word would make a big difference. For example, "secrete" and "release" are different. In Chinese they all mean "放", but in English they refer to different ways of "放". "Secrete" means producing and discharging a substance, whereas "release" refers to the discharging process yet does not emphasise where the substance is produced.*

The teacher's awareness of learners' language-related knowledge also includes their understanding of learners' difficulties in expressing scientific ideas in English. While the teaching of subject matter is often isolated from the teaching of language skills in content subject classrooms, the challenges for students are not only the conceptual understanding of subject contents, but also the representational demand for using appropriate language to articulate specific ideas (Seah, Clarke, & Hart, 2015). Besides, the L2 linguistic competency of students is also of considerable concerns in CLIL classrooms.

Considering both scientific language and English language as essential objectives in CLIL learning, Teacher A and B acknowledge that students encounter the duel challenges to speak up in science classes: first, the linguistic difficulty, namely to express scientific ideas in English, due to their limited English vocabulary and grammatical structures; second, the representational difficulty of describing and explaining scientific phenomena and causative relations by using accurate scientific terms:

4. (TA) *I often find it difficult for them to express scientific ideas in English due to their limited vocabulary. Even if they have rememorised the key terms in English, but how to link them together becomes a considerable difficulty, such as expressing the causative links between X and Y.*
5. (TB) *It is fine for them to use their own words to describe the conversion of the three states of matter. But they will have problems if they are required to employ scientific terms. They may not be able to describe the scientific phenomena correctly. Or when they try to describe, they cannot satisfactorily state the causative relations.*

The comments by both Teacher A and B suggest that students' linguistic and representational difficulty can be intertwined. Moreover, these two aspects would exacerbate their difficulty of grasping content knowledge (or vice versa). This manifests the challenges of learning and teaching in the CLIL programs.

4.1.3 Teachers' understanding of the impact of MOI policy on learner's motivation and learning outcomes

The language policy of the school that participated in this study states that the school adopts English as MOI and yet attaches importance to bilingual and trilingual education among their students. As most students from the examined two classes graduated from CMI primary schools, to adapt to the EMI environment amounts to a challenge for them. Teacher A and B are aware that they are supposed to teach content subjects in English in accordance with the requirements of the Education Bureau. Students are encouraged, but not strictly required, to speak in English. In practice, teachers have the flexibility to use English and Cantonese. As shown by the quotes 6–8, both teachers comment on the positive aspects of English as MOI, which helps students pave the way for learning science-related subjects in English in senior forms and at the university level. However, they raise concerns about the negativity of implementing the EMI policy in a rigid fashion, which leads inevitably to a slower teaching pace, reduced efficacy in giving instructions, and compromising students' scientific interests and willingness to speak in class.

6. (TA) *This is an EMI school. In the long term, studying science in English from Form 1 can help them lay a good foundation.*
7. (TA) *To be honest, for junior form students, they used Chinese in their primary school and it is natural for them to use the language they feel comfortable with when they complete a specific group task, especially when they need to communicate and discuss with other students to solve a problem. I understand that. But are they not supposed to speak in English for this is an EMI school? ... Students insert Chinese words and phrases during the group work... They use English when speaking to the teacher.*
8. (TB) *The school emphases the use of English. I try my best to teach in English, except when it comes to things like safety instructions, which I will use Chinse to remind students.*

Both Teacher A and B are aware that the choice of language would affect students' **motivation** of learning. They believe that forcing students to use English in science class would diminish students' learning interests (see quotes 9, 10).

9. (TA) *It takes time to get familiar with learning science in English. ... If they are required to learn this subject in 100% English and no Chinese at all, their learning interests would be compromised and the learning pace would be slowed down.*
10. (TB) *Students respond more actively if I use Cantonese to ask questions in class. I am sure. If they have to use English, their enthusiasm may be compromised. They can use Chinese freely to elaborate what they mean.*

Both teachers are aware that the choice of language would affect students' **learning outcomes.** Teacher B acknowledges that using English as MOI can improve students' English language skills (see quote 11). His terse response also demonstrates his belief that the exposure to an English learning environment would generate positive outcome of students' English language level.

11. (TB) *Their English would be better. This is quite obvious.*

However, Teacher A raises her concerns (see quotes 12, 13) about the choice of language would exert negative impact upon students' exam results of the science subject, because it remains challenging, especially for students who are weak in comprehension, to take written tests in English.

12. (TA) *If the exam questions contain a long descriptive paragraph, students may not follow easily. Some students are weak in comprehension. ... English language considerably affects their performance in the exams.*
13. (TA) *It may be quicker [if students use Chinese] to understand the scientific knowledge.*

4.2 Conceptual scaffoldings

The scaffolding strategies in the observed sequence of lessons can be classified into six categories:

1. **Mediation**: The teacher recasts to change the registers, with a purpose of helping students make changes between the everyday, oral language and the specialist, academic language. Mediation between everyday and specialist language is a two-way process. Registers and modes of utterance can be shifted in accordance with the target of the exercises.
2. **Probing for Expansion**: The teacher usually asks why/what/how questions to solicit higher order thinking among students and expand the conversation. Students can further elaborate, clarify and cite personal or everyday examples to resolve problems.

3. **Translating**: The teacher makes use of Chinese (Cantonese), which is students' mother tongue. Sometimes, the teacher would translate directly a term into Chinese and even further explain it in Chinese, if it is deemed difficult for students. At times, the teacher would speak the term in Cantonese and ask students to translate it into English.
4. **Evoking Discussions among Students**: The teacher organises the group discussions among students. With a clearly-defined goal of completing a certain task, students' discussion can be productive through reciprocal scaffolding.
5. **Encouraging Students' Self-Scaffolding through Presenting and Re-presenting**: The teacher signals to students the need to reformulate their utterances. In doing so, the teacher offers a chance for students to articulate and reformulate their expression of ideas.
6. **Withholding the Scaffolding**: The teacher stimulates students' scientific interests and self-scaffolding by withholding the scaffolding.

The following excerpts present one or more scaffolding strategies used by Teacher A and B.

In Episode 1 Teacher A organises group discussion and asks students to give examples of different frame structures. In the Initiation-Response-Feedback (IRF) exchange between Teacher A and Group 6, the teacher keeps re-initiating questions about the rationale behind the Japanese model (see line 4, 8). In probing students' higher-order thinking by asking the "why" questions and giving them time to respond, the teacher opens up the learning space. In the second round of Re-Initiation, Harry articulates a half sentence "Because it is…" (line 9). Maybe he has already got the idea, but he fails to come up with the word "earthquake". James then gave the right word (see line 10), which can be seen as a reciprocal scaffolding. When Teacher A revisits the Japanese model in front of the whole class, Harry shouts out the correct answer. It is noteworthy that Harry modifies the sentence from the subject-led sentence structure "Because they always have", which appears somewhat colloquial, to a "there-be" structure to articulate a factual statement (line 14). This linguistic shift is enabled through Harry's self-scaffolding agency. From these two-connected dialogues, it is evident that Teacher A attempts to encourage students to have classroom talk, and allows them to reason, to collaborate and to self-correct in the process of presenting and re-presenting ideas.

Episode 2 is a conversation between Teacher A and Group 3 during the task of searching online information related to the frame structures. Group 3 are browsing through a website. To answer the teacher's question, Lili intends to read aloud a sentence about the function of frame structures but encounters difficulty (see line 2). Knowing that Lili may not understand what "loading" means, Teacher A

Episode 1.

Line	T/S	Moves	Classroom talk	Codes of conceptual scaffoldings
			Teacher-Guided Group Discussion	
1	TA	I	(20:13) [to Group 6] Give one or two examples [of frame structures] in your group form. But then you try to think why they got different names? You can look at this.	*Evoking Discussion among Students* *Probing for Expansion*
2	Charles	R	Japanese model.	
3	TA	F	Nice, yes. Japanese got good designed structures.	
4	TA	Re-I	This is because what reasons? They must develop good frame structure. Why?	*Probing for Expansion*
5	Ben	R	Because of the cold weather?	
6	TA	F	Not really.	
7	Harry	R	Keep the building safe.	
8	TA	Re-I	Why?	*Probing for Expansion*
9	Harry	R	Because it is…	
10	James	R	Earthquake.	*reciprocal scaffolding*
			Later in this lesson the teacher recaps and revisits the above group conversation in front of the whole class	
11	TA	I	(40:20) [to the class] Actually the use of frame structure is to support the weight. Very important. Okay, prevent the building from deformation. "變形". And then, Charles just gave me a very, a country which is good in designing the frame structure. Can you show to your classmates? [to Charles] Which country?	*Translating*
12	Charles	R	Japan.	
13	TA	I	Why? Why Japan need to have a good frame structure?	*Probing for Expansion*
14	Harry	R	Because they always have…Because there are many earthquakes in Japan.	*Encouraging Students' Self-Scaffolding through Presenting and Re-Presenting*

Line	T/S	Moves	Classroom talk	Codes of conceptual scaffoldings
15	TA	F	Yes, therefore they must have some good frame structures to prevent the buildings from deformation. Okay, to make it safe and stable in some critical, very dangerous conditions, just like the earthquake.	

Episode 2.

Line	T/S	Moves	Classroom talk	Codes of conceptual scaffoldings
			Group Work: Students search on-line information to answer questions	
1	TA	I	(18:03) [to Group 3] Could you tell me what is the use of frame structures?	
2	Lili	R	[Lili looks at the tablet] It is used to overcome the large moments developing…	
3	TA	I	Where, where, which one?	
4	Lili	R	This one.	
5	TA	F	"Developing." "Moments." Ah yes, actually, means, this sentence is quite difficult for Form 1 students. This means, "loading", "loading" means the weight. Weight means the mass, just like the kg, the gram, or sometimes in VC called the Newton. Okay, so actually it is to support, support the weight.	Mediation
6	Lili	R	Use the structures to support.	
7	TA	F	Yes, very good. You got the answer already and then you can type [in the form]. Okay.	

recasts to shift the register by using a series of relatively more commonplace concepts: loading → weight → mass (kg; g) → Newton (see line 5). Upon the teacher's exposition, Lili is able to reformulate her answer, instead of copying the original text with a limited degree of comprehension (see line 6). Lili's answer marks an important moment of practising trans-languaging, in which Teacher A's mediation helps her bridge the gap between the specialist, academic language and the one that is more familiar, everyday, and accessible. In this process, the teacher's mediation as a scaffolding is transformed into the student's learning of language across different registers.

Episode 3.

Line	T/S	Moves	Classroom talk	Codes of conceptual scaffoldings
1	TB	I	[To the whole class] Some students, your design, I saw it, is trying to hold it from sideway. Trying to hold the container from the side. [Non-verbal: TB holding a container from sideways with both hands] You have to think, whether hold it by side can provide more supporting force, or you hold it by this can provide more supporting force. [Non-verbal: TB placing the container on his palm] Okay, trying to estimate, Okay? You think it yourself. Of course, I won't tell you the answer.	*Withholding the Scaffolding*
2	S	R	[Non-verbal: returning to their group work]	

As Episode 3 shows, Teacher B draws all students' attention in the middle of their hands-on activity. He inspires students to think about the structural capacity of carrying weight, yet without giving away the answer. By withholding the further scaffolding, the teacher also makes his intentions explicit, namely to stimulate the students' interests and independent thinking to explore the answer. In doing so, the teacher encourages self-scaffolding.

5. Discussion

Teacher language awareness constitutes the teacher's self-reflective knowledge about language input and output in the classroom. This study specifies three salient aspects of teachers' knowledge of how language mediates learning in a specific CLIL context. These aspects address the distinct and interrelated roles of language: (1) as a condition of learning, (2) as an object of learning, and (3) as an institutional and contextual factor that affects learning. Teachers' awareness of the relationship between language and learning is of vital importance, impacting not only on their own using and analysing language, but also on their ways of teaching language in the CLIL classroom. This study calls attention to the very act of learning and learners' needs in the conceptualisation of teachers' language awareness. The shift from teachers to learners expands the framework of TLA and provides insight into what TLA should embody and how it can improve the flow of language in the CLIL classroom.

By orienting the focus of TLA on learning, this study sheds light on the transformation of teachers' implicit knowledge of language into the explicit awareness of how to procedualise language teaching. Although teachers' choices of scaffolding strategies during the talk-in-interaction are contingent upon various factors in practice (Andrews, 2007), the intervention of the teacher's language awareness in the act of scaffolding is discernible. TLA underlines teachers' decision of how to provide conceptual scaffolding related to language learning in CLIL contexts and whether to withhold scaffolding in a strategic manner.

Meanwhile, scaffolding choices also reflect the teachers' awareness of language use in science classrooms. The strategies (1) *mediation* (between colloquial, common language and more specialist, academic language) and (2) *probing for expansion of learning space* are the outcomes of teachers' understanding of learners' language levels, especially related to the content subject. The strategy (3) *translating* addresses teachers' awareness about students' linguistic difficulties under the EMI language policy. These scaffolding strategies align with the students' concerns and the CLIL learning objectives as understood by teachers. In other words, the teachers' awareness of language from the perspectives of learners and learning can direct teachers to optimise how, when and to whom scaffolding is provided. In this sense, the reflective relationship between TLA and scaffolding strategies revolves around the needs of learners and learning.

Notably the strategies (4) *evoking students' discussions*, (4) *encouraging students' reformulation of expressions* and (6) *withholding scaffolding*, manifest teachers' meta-linguistic awareness or consciousness of learning through language, highlighting the necessity of students' engagement in classroom talk and the importance of encouraging students' reciprocal scaffolding and self-scaffolding agency. In doing so, students can be empowered to be competent learners and speakers of the language of science. As Holton and Clarke (2006) remind us, to promote a better construction of knowledge and independent learning for the long-term interest of learners, it is meaningful to realise "the progressive devolution of the role of scaffolding agent from teacher to learner" (p. 141). The TLA-filtered and scaffolded interactions thereby promote the use of language not only for pedagogical purposes but as a cognitive learning tool for students.

It is necessary to add that a scaffolding strategy can be the operation of one or more aspects of TLA in relation to learning of and through language. Although the reflective relationships between scaffolding and TLA can be criss-cross, it is vital to place great emphasis on the teachers' knowledge of learners and learning process in the operation of TLA to facilitate scaffolded dialogues in the CLIL classroom.

6. Conclusion

This study suggests that teachers' awareness of language in CLIL contexts should incorporate the thinking of language in learning and language for learners. Raising content teachers' TLA from the perspectives of learners and learning can effectively enhance teachers' scaffolding role in CLIL science classrooms. There is a definite (if not deterministic) relationship between teachers' awareness of language learning and their scaffolding strategies, especially when both are geared to the needs of learners and learning. This connection between TLA and scaffolding provides a useful guide for teachers and teacher educators to concretise teachers' awareness of language use in the attempts to engage and enable their students in the classroom talk-in-interaction. Further research is required to investigate the ways in which students' needs affect and modify teachers' awareness of language in the CLIL classroom.

Acknowledgements

This research project was supported by the funding from the Standing Committee on Language Education and Research (SCOLAR). We thank the two teachers and students for their participation, and the whole research team for data collection and analysis. We also thank two anonymous reviewers for their suggestions.

References

Andrews, S. (2007). *Teacher language awareness*. Cambridge: Cambridge University Press. https://doi.org/10.1017/CBO9780511497643
Andrews, S., & Lin, A. M. Y. (2018). Language awareness and teacher development. In P. Garrett & J. M. Cots (Eds.), *The Routledge handbook of language awareness* (pp. 57–74). London: Routledge.
Andrews, S., & Svalberg, A. M.-L. (2017). Teacher language awareness. In J. Cenoz, D. Gorter, & S. May (Eds.), *Language awareness and multilingualism*. (pp. 219–231). New York, NY: Springer.
Chan, J. Y. H. (2016). The fine-tuning medium-of-instruction policy in Hong Kong: A case study of the changing school-based test papers in science subjects. *Education Journal*, 44(1), 159–193.
Choi, P. K. (2003). "The best students will learn English": Ultra-utilitarianism and linguistic imperialism in education in post-1997 Hong Kong. *Journal of Education Policy*, 18(6), 673–694. https://doi.org/10.1080/0268093032000145917
Coyle, D., Hood, P., & Marsh, D. (2010). *CLIL: Content and language integrated learning*. Cambridge: Cambridge University Press.

Edge, J. (1988). Applying linguistics in English language teacher training for speakers of other languages. *ELT Journal*, 42(1), 9–13. https://doi.org/10.1093/elt/42.1.9

Education Bureau. (2009). *Fine-tuning the medium of instruction for secondary schools*. Hong Kong SAR: Education Bureau.

Evans, S. (2013). The long march to biliteracy and trilingualism: Language policy in Hong Kong education since the handover. *Annual Review of Applied Linguistics*, 33, 302–324. https://doi.org/10.1017/S0267190513000019

Fung, D., & Yip, V. (2014). The effects of the medium of instruction in certificate-level physics on achievement and motivation to learn. *Journal of Research in Science Teaching*, 51(10), 1219–1245. https://doi.org/10.1002/tea.21174

García, O. (2009). *Bilingual education in the 21st century: A global perspective*. Malden, MA: Wiley-Blackwell.

Gibbons, P. (2003). Mediating language learning: Teacher interactions with ESL students in a content-based classroom. *TESOL Quarterly*, 37(2), 247–273. https://doi.org/10.2307/3588504

Graham, K. M., Choi, Y., Davoodi, A., Razmeh, S., & Dixon, L. Q. (2018). Language and content outcomes of CLIL and EMI: A systematic review. *Latin American Journal of Content and Language Integrated Learning*, 11(1), 19–37. https://doi.org/10.5294/laclil.2018.11.1.2

Halliday, M. A. K. (1993). Towards a language-based theory of learning. *Linguistics and Education*, 5, 93–116. https://doi.org/10.1016/0898-5898(93)90026-7

Hoare, P. (2003). Effective teaching of science through English in Hong Kong secondary schools. (Unpublished doctoral thesis). The University of Hong Kong, Hong Kong. https://doi.org/10.5353/th_b2976829

Holton, D., & Clarke, D. (2006). Scaffolding and metacognition. *International Journal of Mathematical Education in Science and Technology*, 37(2), 127–143. https://doi.org/10.1080/00207390500285818

Koopman, G. J., Skeet, J., & de Graaff, R. (2014). Exploring content teachers' knowledge of language pedagogy: A report on a small-scale research project in a Dutch CLIL context. *The Language Learning Journal*, 42 (2), 123–136. https://doi.org/10.1080/09571736.2014.889974

Lasagabaster, D., & Sierra, J. M. (2016). Immersion and CLIL in English: More differences than similarities. *ELT Journal*, 64(4), 367–375. https://doi.org/10.1093/elt/ccp082

Lindahl, K. M. (2013). Exploring an invisible medium: Teacher language awareness among preservice educators of English language learners (Unpublished doctoral thesis). University of Utah, Salt Lake City, UT.

Lindahl, K. M. (2016). Teacher language awareness among pre-service K-12 educators of English language learners. In J. Crandall & M. Christison (Eds.), *Teacher education and professional development in TESOL: Global perspectives* (pp. 127–140). New York, NY: Routledge. https://doi.org/10.4324/9781315641263-8

Lindahl, K. M., Baecher, L., & Tomas, Z. (2013). Teacher language awareness in content-based activity design. *Journal of Immersion and Content-Based Language*, 1(2), 198–225. https://doi.org/10.1075/jicb.1.2.03lin

Lindahl, K. M., & Watkins, N. (2015). Creating a culture of language awareness in content-based contexts. *TESOL Journal*, 6(4), 777–789. https://doi.org/10.1002/tesj.223

Liyanage, I., & Bartlett, B. J. (2010). From autopsy to biopsy: A metacognitive view of lesson planning and teacher trainees in ELT. *Teaching and Teacher Education*, 26, 1362–1371. https://doi.org/10.1016/j.tate.2010.03.006

Lo, Y. Y. (2019). Development of the beliefs and language awareness of content subject teachers in CLIL: Does professional development help? *International Journal of Bilingual Education and Bilingualism*, 22(7), 818–832. https://doi.org/10.1080/13670050.2017.1318821

Lyster, R. (2007). *Learning and teaching languages through content: A counterbalanced approach.* London: John Benjamins Publishing. https://doi.org/10.1075/lllt.18

Merino, J. A., & Lasagabaster, D. (2018). The effect of content and language integrated learning programmes' intensity on English proficiency: A longitudinal study. *International Journal of Applied Linguistics*, 28(1), 18–30. https://doi.org/10.1111/ijal.12177

Morton, T. (2018). Reconceptualizing and describing teachers' knowledge of language for content and language integrated learning (CLIL). *International Journal of Bilingual Education and Bilingualism*, 21(3), 275–286. https://doi.org/10.1080/13670050.2017.1383352

Regalla, M. (2012). Language objectives: More than just vocabulary. *TESOL Journal*, 3(2), 210–230. https://doi.org/10.1002/tesj.15

Seah, L. H., Clarke, D., & Hart, C. (2015). Understanding middle school students' difficulties in explaining density differences from a language perspective. *International Journal of Science Education*, 37(14), 2386–2409. https://doi.org/10.1080/09500693.2015.1080879

Shulman, L. S. (1987). Knowledge and teaching: Foundations of the new reform. *Harvard Educational Review*, 57(1), 1–22. https://doi.org/10.17763/haer.57.1.j463w79r56455411

Sinclair, J., & Coulthard, M. (1975). *Towards an analysis of discourse: The English used by teachers and pupils.* London: Oxford University Press.

Tan, M. (2011). Mathematics and science teachers' beliefs and practices regarding the teaching of language in content learning. *Language Teaching Research*, 15(3): 325–342. https://doi.org/10.1177/1362168811401153

Thornbury, S. (1997). *About language.* Cambridge: Cambridge University Press.

Vygotsky, L. S. (1978). *Mind in society: The development of higher psychological functions.* Cambridge, MA: Harvard University Press.

Wright, T., & Bolitho, R. (1993). Language awareness: A missing link in language teacher education? *ELT Journal*, 47(4), 292–304. https://doi.org/10.1093/elt/47.4.292

Supporting students' content learning in Biology through teachers' use of classroom talk drawing on concept sketches

Caroline Ho, June Kwai Yeok Wong and
Natasha Anne Rappa
English Language Institute of Singapore, Ministry of Education
Academy of Singapore Teachers, Ministry of Education
Murdoch University

This article examines teachers' attempts to enhance students' content learning in Biology through the use of talk centred on concept sketches. Of specific interest is how teachers provide scaffolding through purposeful classroom discourse (Lemke, 1990) with the use of talk moves (Chapin, O'Connor, & Anderson, 2013), drawing on concept sketches (Johnson & Reynolds, 2005) annotated by students. Informed by socioconstructivist (Vygotsky, 1978/86) perspectives and grounded in multimodal literacy (Kress & van Leeuwen, 2001) underpinnings, the study acknowledges the teacher's role in productive classroom discussions to guide students' thinking and facilitate meaning-making. Qualitative analysis of classroom discourse illustrates how teachers' classroom talk can scaffold and address the gaps in students' learning. Pedagogical implications are discussed.

Keywords: classroom discourse, concept sketch, multimodal literacy, content and language integrated learning (CLIL), science teaching, Biology, teacher questioning, talk moves

1. Introduction

Biology as a visual discipline (Bell, 2014) poses challenges to students as it involves 'meaning making through multiple modalities' (Jaipal, 2009, p. 48). Teaching Biology becomes equally complex as a process (Jaipal, 2009) not only for its multimodal considerations (Prinou & Halkia, 2003) but also the 'challenges associated with learning the language of science along with its modes of representation' (Bennett, 2011, p. 34). Teachers of Biology at the pre-university level in a Singapore junior college observed that their students faced difficulties in understanding abstract concepts and making sense of biological processes. Among the specific communication skills required of Biology students are the following pertaining to demonstrating science inquiry skills as stated in the Ministry of Education (Curriculum Planning and Development Division, 2016) and assessment syllabus (MOE & UCLES, 2015): 'use appropriate models to explain concepts, solve problems and make predictions' and 'communicate scientific findings and information using appropriate language and terminology' (MOE & UCLES, 2015, p. 4). The appropriate scientific vocabulary demanded of students include terminology and conventions (including symbols, quantities and units) (MOE & UCLES, 2015, p. 7) that are relevant for explaining scientific phenomena. This study examines Biology teachers mediating content learning conveyed visually and in text through purposeful classroom talk (Lemke, 1990; Lin, 2016). Attention focuses on teachers' use of classroom talk to help students make sense of key processes and establish conceptual links through the use of students' concept sketches (Johnson & Reynolds, 2005).

2. Theoretical underpinnings

The principal theoretical framework underlying this study is social constructivism grounded in Vygotskyan (1978, 1986) influence, with a focus on how knowledge is constructed in the social context of the classroom through language and other semiotic means. Science research has foregrounded science as very much about 'making new things visible, or familiar things visible in new ways' (Wise, 2006, p. 75) where 'images form both what and how we know' (Wise, 2006, p. 82). To facilitate making visible what is required in science, scaffolding acknowledged as assisted performance that guides a learner to complete a task and develop the capacity to manage learning independently (Scott, 1997) is critical in supporting students' learning. It enables students to accomplish tasks which they would not have been able to manage on their own and help them complete such tasks competently (Hogan & Pressley, 1997; Michell & Sharpe, 2005).

In reading, writing and talking about discipline-specific content in Science, there is a need to consider how students can be supported, given the multimodal nature of knowledge representations and meaning-making (Kress, Jewitt, Ogborn, & Tsatsarelis, 2001; Kress & van Leeuwen, 2001), and to consider how the modes, that is, different semiotic resources for representation and communication (Kress, 2010; Kress & van Leeuwen, 2001), contribute to the meaning-making process. The need for students to actively construct representations to become competent in scientific practices and learn through participating in the reasoning processes of science (Ford & Forman, 2006) has been highlighted in science education. Students have to 'translate, integrate and reinterpret meanings across verbal, visual and mathematical representations, as well as connect these modes to earlier experiences of science activity' (Prain & Hand, 2016, p. 4). Bransford and Schwartz (1999) have pointed to the need to re-conceptualize the learning gains and acknowledge the potential for transfer of learning through students generating their own representations (Tytler & Hubber, 2016, p. 160).

Semiotic resources for representation and communication in science include not just visual representations but also language. Language is acknowledged to be important not only as the instrument of interactions (cultural tool) but also as an instrument of verbal thought (cognitive tool) (Vygotsky, 1986) to facilitate knowledge construction. Of specific interest in this study is providing scaffolding via classroom talk (Lin, 2016, p. 105) with language recognized as the key meaning-making resource for constructing content (Lemke, 1990; Lin, 2016) and as a symbolic tool (Walsh, 2013) to clarify and make sense of new knowledge through oral communication. How teachers 'scaffold communicating in science both in spoken texts and written texts, and in appropriate spoken genres and written genres' (Lin, 2016, p. 61) is critical as language offers a 'means by which one comes to know what one does not know' (Swain & Lapkin, 2013, p. 105) through mediating thought (Vygotsky, 1978, 1986; Swain & Lapkin, 2013). Greater attention to scaffolding in instructional strategies in multimodal contexts through classroom talk underscores language as a critical meaning-making resource for content learning evident in content-based instruction and content and language integrated learning (CLIL) contexts (Dalton-Puffer, 2013).

3. Review of studies in the field

The select review focuses on studies on concept sketch before attention turns to classroom talk in learning science contexts.

3.1 Concept sketch in the field of science

In biology, the visual representation of processes and concepts enables students to make sense of and see the relationships between concepts. Student-generated visual representations are recognised to be valuable as a powerful tool for thinking and communicating (Roam, 2008; Temple, 1994). Visual representations are integral to the practice of science, and facilitate the generation of hypotheses, data interpretation and communication of results (Ainsworth, Prain, & Tytler, 2011; Schwarz et al., 2009). Guiding students in their content learning through visual representations is recognized to 'reduce cognitive load, enhance representation of relationships among complex constructs' (O'Donnell, Dansereau, & Hall, 2002, p. 72).

One form of visual representation is concept sketches which are 'sketches or diagrams that are concisely annotated with short statements that describe the processes, concepts and interrelationships shown in the sketch' (Reynolds & Tewsbury, 2005, p. 1). Similar to annotated visual organizers with concise labels, they enable learners to identify the features of the concepts illustrated; describe and explain the processes involved; and 'characterise the relationships between features and processes' (Johnson & Reynolds, 2005, p. 86) or illustrate hierarchies between related concepts (Novak, 1998). Research shows that concept sketches help students construct and organize knowledge and learn more than those who do not construct concept sketches (Esiobu & Soyibo, 1995; Novak & Wandersee, 1991).

3.2 Classroom talk in science

Studies have shown that questions and prompts that teachers use to structure classroom interactions play a significant role in scaffolding (Kawalkar & Vijapurkar, 2013). Indeed, the types of questions teachers ask and how they are asked can, 'to some extent, influence the type of cognitive processes that students engage in as they grapple with the process of constructing scientific knowledge' (Chin, 2007, p. 816). Scott (1998) in reviewing studies on how meanings are developed in the science classroom, believes that students' dialogic discourse forms the basis of active, analytic, individual thought through questioning and relating ideas. Lemke (1990) sees talking science as 'doing science through the medium of language' (p. ix) whereby specific ways of 'talking, reasoning, observing, analysing and writing' (p. xi) as social processes are constructed and valued by members within the scientific community. The semantic resources of language used in talking science facilitate communicating conceptual relationships though classroom dialogue. Dawes' (2004) work with younger learners on scientific concept development involved structured activity mediated through oral language. Findings indicated that allowing students to engage with each other provides 'crucial practice in the use of new vocabulary in context' and 'can help learners to exchange ideas, to have access to different perspectives and to make meaning together' (p. 693). Chin's (2007) study of teachers' questioning to scaffold student thinking and construct scientific knowledge showed questions could be framed through specific approaches to support student learning and foster productive student responses. Chia, Tay, Ho, Ho, & Lee's (2014) research into Singapore teachers supporting Chemistry students in constructing scientific explanations focused on scaffolding via classroom talk to make explicit the literacy demands required. The study indicated that 'active student engagement facilitated by concrete support scaffolds in various modalities (visual, textual, oral) and resources' (p. 33) can engage students in the process of learning, and 'contribute to the dialogic process of knowledge co-construction and purposeful meaning-making' (p. 33).

Most documented studies examined classroom talk and multimodal literacy as isolated practices. Recent work in multimodality and classroom talk has attempted to explore the relationship between patterns of classroom talk and teachers' modes of representation (Tang, 2016). The study indicated shifts in classroom communication patterns depending on the modes of representation used-enactive (action based), iconic (image based) or symbolic (language based). Tang (2016) emphasized 'the importance of considering the role of representation and its transformation process within the context of the classroom communicative approach (e.g., dialogic or authoritative)' (Tang, 2016, p. 2091) and how the shift from one communicative approach to another is mediated through the use of representations. As Tang (2016) noted in the study's limitation, more research in this area is needed in other branches of science including biology. While Tang (2016) examined the shifts in the communicative approach through representations, this paper focuses specifically on the discourse moves of teachers to help students in content learning based on their annotated concept sketch.

Overall, there remains a dearth of investigations with documented research in science learning contexts which examine meaningful classroom talk drawing on students' visual representations as a means of scaffolding the learning of science. This study, in supporting students within a specific pedagogic context in Biology to fulfil targeted curricular goals, fills a gap in the current local research and pedagogic contexts. It reinforces what research has highlighted where 'learning is realised through the interaction between visual, actional and linguistic communication (i.e., learning is multimodal) and involves the transformation of information across different communicative systems ('modes'), e.g., from speech to image' (Jewitt et al., 2001, p. 5). Supporting students' content learning through foregrounding how teachers' classroom talk (speech) drawing on students' concept sketch (image) with written annotations (text) can raise students' awareness and focus attention on the specifics and details required in the specific context.

4. Research focus

The overall objective of this study was to examine teacher-facilitated classroom discussions drawing on students' concept sketches in Biology. Specifically, the study was guided by the following research question:

How can teachers use talk to support the students' learning in Biology through concept sketches?

5. Background setting and subjects

The work drew on a research collaboration between the English Language Institute of Singapore and a college through the support provided under the Whole School Approach to Effective Communication in English (WSA-EC), a strategic Ministry of Education (MOE) initiative aimed at developing teachers' ability to communicate subject knowledge more clearly and effectively to support students' learning (English Language Institute of Singapore, 2011). Grade 11 (sixteen to seventeen years old) male and female students from two classes of nineteen and twenty two students respectively were involved. The majority of the students were from a predominantly middle to lower socio-economic background with average to less than average ability in Biology. For students in multi-ethnic Singapore, English is not necessarily the dominant home language for many. Based on official records, English is the home language for 61% of Chinese children, 54% of Indian children, 36% of Malay children between the ages of 5 and 14 (Singapore Department of Statistics, 2015). From 1987, English has been the medium of instruction for all subjects, except for the mother tongue languages (The Straits Times, 1983). This study is particularly relevant to learning in content and language integrated learning contexts where English is the medium of instruction. Teachers between three and five years of pre-university science teaching experience were involved.

6. Methodology

This study was an instantiation of 'design research' (Collins, Joseph, & Bielaczyc, 2004) intervention involving close collaboration between researchers and teachers in planning and co-developing the instructional materials used. The study was contextualised within the regular college science curriculum with a focus on the topic of chemiosmosis, a process in cellular respiration and photosynthesis in Biology which students have difficulty understanding. The data used in this paper are drawn from a fifty-minute lesson which reviewed concepts related to the topic of chemiosmosis. This lesson served to scaffold and consolidate students' content learning which would help them subsequently in their written responses to questions on the targeted topic. Specifically, in this article, attention centred on content learning related to the 'stalked particle', its orientation and component features through student-generated concept sketches and the discussion around this aspect, given that this is the most prominent feature of the mitochondria.

Students were given the task of labelling the different parts of the mitochondria before they attempted to draw the stalked particle on the diagram. Teachers monitored students' attempts as they drew and identified specific students with inaccurate representations to share their work with their peers. Teachers engaged students in classroom discussion through the use of talk moves at specific segments of the lesson. Teachers earlier worked with the researchers in their own professional learning on integrating the use of the question prompts in their content teaching. The use of specific prompts for particular purposes (Appendix B) aimed to elicit students' thinking and sharpen students' understanding of the relevant content based on their concept sketch.

The primary data source for the study reported in this paper was transcripts of classroom discourse unedited from video-recorded lessons. Other data sources comprised teachers' instructional materials, students' written tasks, teachers' focus group discussions and students' surveys. Subjects' names in transcripts were replaced by pseudonyms and 'Students' indicate a chorus response from several students. Qualitative research methodology was adopted with the researchers examining classroom talk based on discussions of students' concept sketches. The focus was on identifying the range of 'talk moves', that is, 'strategic ways of asking questions and inviting participation in classroom conversations' (Chapin, O'Connor, & Anderson, 2013, p. 11) adopted for specific purposes, and in categorising the types identified. The work was guided by a framework developed by the English Language Institute of Singapore which was adapted from earlier work in scaffolding academic classroom discussions (Michaels & O'Connor, 2012; Zwiers & Crawford, 2011) drawing on dialogic teaching principles (Alexander, 2008) for facilitating productive academic classroom discussion for specific purposes. The analysis of talk moves was examined in relation to the relevant parts of the student-generated concept sketch in focus. The categorisation of talk moves was carried out independently first by the researchers who then cross-checked with each other for consistency in classification.

The teachers planned with a view to integrate talk moves into lessons (Appendix A) based on students' concept sketches to help students explicitly verbalize their thinking. This is aimed at sharpening accuracy and precision in students' reasoning, and in guiding students to observe, reflect and make logical links during class discussion. Students were tasked to draw and write their response to questions raised by the teacher and classmates. Activity sheets guiding students in the representation of their concept sketches were provided.

7. Examination of classroom talk based on concept sketch

The illustrative data in this paper draw out students' misconception and show how this was addressed. The teacher required students to draw arrows to indicate the direction of the flow of hydrogen ions (H+). Using a visualizer, the teacher presented a student's work displaying H+ concentrations where the drawing of the H+ was incorrect. The teacher facilitated a classroom discussion through the use of specific talk moves drawing on the student's concept sketch (Figure 1) to highlight the inaccuracy and surface the misconception.

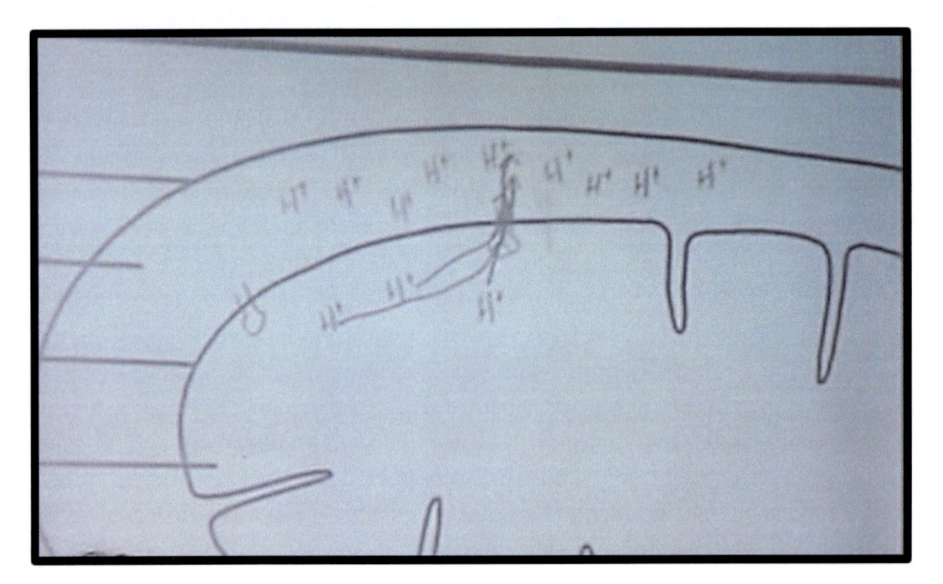

Figure 1. Extract from student's concept sketch showing incorrect direction of flow of hydrogen ions (H+)

The following annotated extracts highlight the specific domains evident in the classroom talk.

7.1 Reformulating with specifics amplified for directionality focus

The teacher elicits a student's response on the movement of H+: 'How will your H+ move?' (line 1) by reformulating to elaborate on the question in different ways with further prompts 'Whatever stations that you can think of', 'can you let me know what is happening here?' (lines 2 and 3) in an attempt to have the student describe specifically the movement of H+. The teacher revoices the student's earlier input (line 6) 'H+ ions are pumped *out of* in line 7 and 'pump *out of* the matrix' (line 4) marked by the preposition of direction 'of' and teacher amplifies with specific location information '*into* the intermembrane space' (line 7) marked by preposition of direction 'into'. This makes explicit the direction involved, that is, a movement out of one area and into the targeted area involved. The teacher's challenge 'Is that all?' (line 7) prompts the student to elaborate on the given response. The teacher follows up with a further probe for reasoning: 'why would your H+ be accumulating' (line 8) through amplifying with location specifics '*in* the intermembrane space' prefaced by the preposition of direction 'in'. This appears to enable the student to think through the H+ attendant use guided by the elicitation for specifics: 'What is it used for? (line 8)'. In line 10, the probe for reasoning 'why do you actually apply your proton gradient?' is reformulated as 'what is the purpose for the flow of electron?'.

Extract 1.

Line/Turn	Speaker	Classroom talk	Talk move
1	Teacher	How your H+ will move?	Elicit specifics
2	Teacher	Whatever stations(??) that you can think of	Reformulate
3	Teacher	so Jane, can you let me know what is happening here?	Reformulate
4	Jane	H+ ions like pump out of the matrix	
5	Teacher	Louder. Cannot hear you.	
6	Jane	H+ ions are pumped out of…	
7	Teacher	H+ ions are pumped out…of the matrix…into the intermembrane space.	Revoice for verification
			Amplify
		Is that…	
		is that all?	Challenge
8	Teacher	So what…why..why would your H+ be accumulating in the intermembrane space?	Probe for reasoning
			Amplify
		What is it used for?	Elicit specifics
9	Student	Proton gradient.	
10	Teacher	Proton gradient.	Revoice
		Why do you actually apply your proton gradient?….	Probe for reasoning
		for flow of electron…	
		what is the purpose for the flow of electron?	Reformulate
		So H+ is being pumped into the intermembrane space for the flow of electron.	Amplify with specifics

7.2 Scaffolding content learning undergirded by purpose

In the next sequence, the teacher's probe for reasoning '*Why* do you actually apply your proton gradient?' (line 10) with the amplified 'for flow of electron' to guide students' thinking is further reformulated as '*what is the purpose* for the flow of electron?'. This is followed by specifics amplification for the purpose or rationale: 'So H+ is being pumped into the intermembrane space for the flow of electron'. This guides students' thinking through the purpose for the process in the given phenomenon.

The prompting for the underlying purpose is evident again in a later sequence: '*Why* do you think that the H+ needs to go out? *What is the purpose?*' (line 37). The teacher's probe for the reason as to why H+ 'needs to go out' seeks to unpack student's content learning for the directionality involved in the given context. This is a point for which the teacher believes bears reiteration to enable students to internalise understanding of the rationale involved. This is again evident in the repeated question '*what is the purpose* of the proton gradient' (line 39).

7.3 Drawing on repeated uptake of students' response to sharpen focus and direct attention

A series of revoice moves in the strategic uptake of students' responses draws on specific parts of students' contributions across a sequence of teacher-student exchange. The teacher's use of specific phrases produced by students sharpens focus of attention on the specifics raised. This serves as a means to direct attention in a more targeted manner. The teacher revoices from the uptake of the student's earlier contribution 'So just now she was saying that your H+ is being pumped into the intermembrane space...*for the flow of electron*' (line 11) as a means to extend further thinking ('what do you think about this?') to elicit student's view on another student's response (line 11). The teacher further revoices the student's previous input 'She said *for the flow of electron*' (line 11).

In the teacher's revoice in line 13 of the student's 'from the flow of electron' (line 12), the prompt for further information in the process in 'what happens?' (line 13) elicits the student's response: 'the H+ come out' (line 14). The teacher builds on the student's input through a consolidated revoice '*Flow of electron*, your *H+ comes out*' (line 15) which integrated the previous two contributions 'flow of electron' (line 12) and 'the H+ come out' (line 14). The one-word prompt 'then?' (line 17) following the revoice in line 17 of the student's earlier contribution 'it gets oxidised' (line 16) seeks to elicit from students the subsequent process involved. This is aimed at prompting students to elaborate further which the student subsequently did with the elaborated response in line 18.

Supporting students' content learning in Biology through classroom talk 95

Extract 2.

Line/Turn	Speaker	Classroom talk	Talk move
11	Teacher	Siew Li, you look surprised. Would you want to say something?	Elicit student's view on other student's response
		So just now she was saying that your H+ is being pumped into the intermembrane space… for the flow of electron.	Revoice for verification
		What do you think about this? You look surprised.	Elicit student's view on other student's response
		She said for the flow of electron.	Revoice for verification
		Just now you "huh mah" so what are you "huh-ing" about?	
12	Siew Li	I thought it's from the flow of electron	
13	Teacher	from the flow of electron, what happens?	Revoice for verification Elicit specifics
14	Siew Li	the H+ comes out..	
15	Teacher	Flow of electron, your H+ comes out	Revoice (consolidated)
16	Siew Li	Like it gets…oxidised?	
17	Teacher	it gets oxidised…ok..	Revoice for verification
		Then?	Elicit specifics
18	Siew Li	the ADH..something like that…then it's oxidised then the H comes out…then remains there..after that it does oxidation then it goes back..then generates	

In the next instance of student uptake, the elicitation (line 21) expands on the previous student's contribution (line 20) with the naming of the 'stalked particle' and the specification marked by the preposition of direction: 'Is it through the stalked particle?' (line 21). This aims at focusing the student's attention on the directionality involved.

As location specifics in the targeted response are critical in understanding the given phenomenon, the teacher's prompt for 'so what is missing from Jane's answer?' (line 23) is followed by the subsequent questioning (line 25) that builds on the student's 'arrow' input (line 24). The specific nature of the 'arrow' and, specifically, its spatial orientation are pursued through the teacher's successive build up of elicitations: 'What arrow? Arrow of what? Arrow of pump particle?.. arrow of H+ into the stalked particle…err… through the stalked particle into the intermembrane' (line 25). The amplification 'through the stalked particle into the intermembrane' (line 25) delineates the directional flow of the targeted object. This is a critical point of reference for students to visualize the exact specifics of the targeted feature.

The repeated elicitation for specifics 'What's wrong with the diagram?' (line 33) reinforces the need for attention to the directionality in the given context. The deliberate specification of the hydrogen ions 'will..go *out..into* the intermembrane space.. *through* electron carrier ..so it will come *back into*' (line 33) directs students' attention to examining the critical aspects of directionality and location specifics involved. The strategic follow up with the pointed question on what is possibly wrong (line 33) is intended to draw out the inaccuracy in the student's representation. The probe for specifics on how it should then be represented: 'so how should you draw this?' (line 35) is reformulated with attention to whether the feature should be pointing out or in: 'So it will go out and then come in via.' (line 35). The arrow as a semiotic resource used by the student shows how directionality is used to realize meaning in visually indicating relationships. The phrases indexing direction 'out of' and 'into the' followed by the resulting phrase specifying direction: 'come in via' are critical to students visualizing the directional flow involved from one point to another. The verbal cloze marked by 'via' following 'then come in' is strategic in drawing out further information. In the subsequent move, the repeated 'what is the purpose?' (lines 37, 39) is focused on eliciting the 'proton gradient' function.

Consolidation moves also feature repeatedly in the cumulative build up which this sequence of turns later develops into: 'just now we were looking at the diagram..err..to show you how the H+...exit to the intermembrane space, correct' (line 39). A further consolidated move 'We were talking about how the H+ will flow… out of your matrix into the intermembrane space. And after that it will come back in via your stalked particle' (line 40) reiterates the directionality earlier emphasized that is critical and is seen to integrate the points raised thus far in the exchange.

Extract 3.

Line/Turn	Speaker	Classroom talk	Talk move	Image
19	Teacher	So how…how does it actually goes back?	Elicit specifics	
20	Siew Li	Through the…		
21	Teacher	Is it through the stalked particle?	Elicit specifics (Amplify)	
22	Siew Li	Yes…		
23	Teacher	so what is missing from Jane's answer?	Elicit specifics	
24	Siew Li	Arrow…. arrow that goes back to the…		
25	Teacher	What arrow? Arrow of what? Arrow of pump particle?.. arrow of H+ into the stalked particle…err… through the stalked particle into the intermembrane	Revoice for verification Revoice with amplification	
26	Jovin	Totally right		
27	Teacher	Totally right?	Revoice for verification	
28	Teacher	Why would you actually say so?	Probe for reasoning	
		Jovin, why do you say this is totally right? Louder, cannot hear you.	Reformulate	
29	Teacher	so what is your opinion? What should be happening here?	Elicit specifics Reformulate	
30	Jovin	it will go out. Not from there. From the electron carrier. I got it wrong…		
31	Teacher	what…what..what..what… what goes out?	Seek clarification	
32	Jovin	H+ should get pumped out.		

Line/Turn	Speaker	Classroom talk	Talk move	Image
33	Teacher	So H+ will… go out…into the intermembrane space…through…electron carrier…so it will come back into the stalked particle.		
		So what's wrong with this diagram? Huh?	Elicit specifics	
		What's wrong with this diagram?	(Repeat) Elicit specifics	
34	Jovin	That one got no…never recycle. Pathway also wrong. Everything wrong		
35	Teacher	so how should you draw this?	Probe for reasoning	
		So it will go out and then come in via..	Reformulate (with verbal cloze)	
36	Jovin	The arrow come in through the stalked particle.		
37	Teacher	through the stalked particle. Why do you think that the H+ needs to go out?	Probe for reasoning	
		What is the purpose?	Reformulate	
38	Class	proton gradient.		
39	Teacher	what is the purpose of the proton gradient?	Reformulate with amplification	
		So then energy… what energy? For you to synthesise ATP. Alright.	Elicit specifics	
		So just now we were looking at the diagram..err..to show you	Consolidate	

Line/Turn	Speaker	Classroom talk	Talk move	Image
		how the H+…exit to the intermembrane space, correct? Let me take a shot of this.	Seek clarification	
40	Teacher	Alright. We were talking about how the H+ will flow… out of your matrix into the intermembrane space. And after that it will come back in via your stalked particle.	Consolidate	

7.4 Sharpening precision in language use

In the following, the teacher specifically emphasizes greater precision in language use for describing the targeted features in the given phenomenon (Figure 2) where students had to draw the stalked particle.

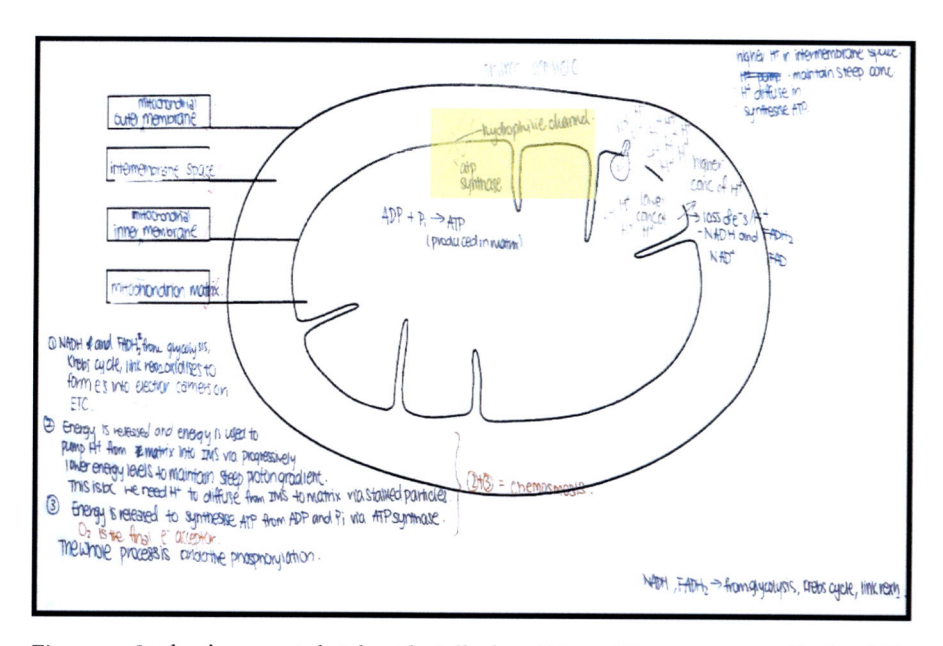

Figure 2. Student's concept sketch with stalked particle and its components ('hydrophilic channel' and 'ATP synthase') highlighted

In the teacher's elicitation of other students' views on the 'difference in terms of the shape' (line 46), this leads to the teacher addressing a student's vague use of language: 'stuff inside' (line 47). The teacher seeks clarification (line 48) from the student by drawing on his imprecise use of the word 'stuff' in the pointed question: 'what do you mean there is stuff inside?' before eliciting from other students the composition of the stalked particle. The teacher is able to draw out from students that the 'stuff inside' actually referred to the specific components 'ATP synthase' (line 51) and 'hydrophilic channel' (line 53). The teacher persists in the line of questioning until the features were identified as 'ATP synthase' and 'hydrophilic channel'. Following this, the teacher then has the students label the part, and only then through the student's response, shows that the bottom of the stalked particle is, in fact, the hydrophilic channel.

Extract 4.

Line/Turn	Speaker	Classroom talk	Talk move	Image
41	Teacher	What part is being projected into the matrix? the bottom of the stalked particle is being projected into the matrix.		
		So the person who drew this, can you clarify why you actually draw it this way?	Probe for reasoning	
42	Teacher	sorry? Louder louder. I can't hear you.		
43	Jasmine	The bottom of the stalked particle contains ATP synthase.		
44	Teacher	so you were saying that this part is actually your ATP enzyme.	Revoice with amplification	
		How come there is an imbalanced size of both parts? So why did you actually draw it this way?	Probe for reasoning	

Supporting students' content learning in Biology through classroom talk 101

Line/Turn	Speaker	Classroom talk	Talk move	Image
45	Teacher	so what does this part represent?	Elicit specifics	
		what does this part represent?		
		what about this part?	Reformulate	
46	Teacher	Just for the main part.. because it's unbalanced…imbalanced	Challenge	
		so why is there such a difference in terms of the shape?	Probe for reasoning	
47	Jasmine	Because there is stuff inside		
48	Teacher	what do you mean there is stuff inside?	Seek clarification	
49	Jasmine	里面有东西 (There is stuff inside)		
50	Teacher	alright. Err..what do you think a stalked particle is made up of?	Elicit specifics	
51	Students	ATP synthase		
52	Teacher	ATP synthase		
53	Students	Hydrophilic channel		
54	Teacher	and the hydrophilic channel. So are we able to identify which is which?	Seek clarification	
55	Students	yes		

Through the teacher-guided classroom discussion, the student is given the opportunity to have his misconception addressed. This leads to his constructing a revised representation (Figure 3) with the directionality of the arrows accurately capturing what is required.

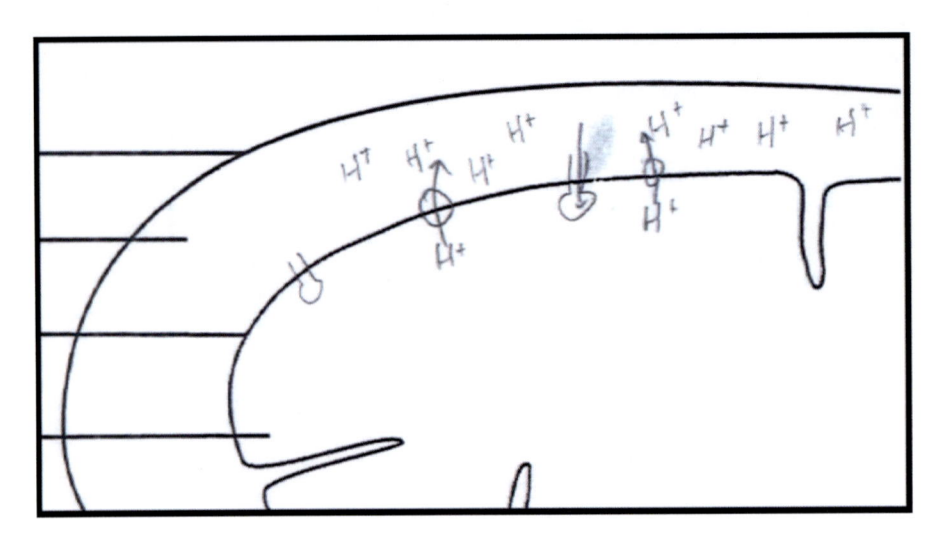

Figure 3. Revised student's concept sketch showing corrected direction of flow of hydrogen ions (H+) with arrows from intermembrane space to matrix through stalked particles and arrows from matrix to intermembrane space

Through a series of reformulation of students' contributions which build on the uptake of students' earlier responses, these strategic moves could possibly serve to help students understand the content specifics relevant to the targeted phenomenon through different ways of elicitation. Revoicing extends beyond a simple repetition in clarifying the student's contribution and giving the student space to verify whether the teacher's restatement is in line with what the student said. As seen, revoicing suggests that students could be helped to re-state with greater clarity or perhaps for students to sharpen specificity on earlier vague responses.

The teacher's moves of amplification with specific details relevant to the points raised at each stage of the exchange suggest that students are guided to concretize what is required. Further, the punctuating of responses with the move that elicits students' understanding of the purpose or rationale at strategic points in the given context is deliberate in co-constructing with students understanding of the 'why' for the 'what'. The teacher's consolidated move which revoices students' earlier responses across several turns appears to help students integrate and pool together the different strands of points raised in the summation of points.

7.5 Limitations in classroom talk

The study also indicated that while teachers' classroom talk appeared to focus on what students' visually illustrated in their concept sketch, teachers did not appear to sufficiently highlight the limitations of students' concept sketch with regard to the written description provided by students. This is critical given the multimodality in science where each mode has distinct functions for meaning-making that complement other modes (Liu, 2018). Collectively, these different modes multiply meaning (Lemke, 1998).

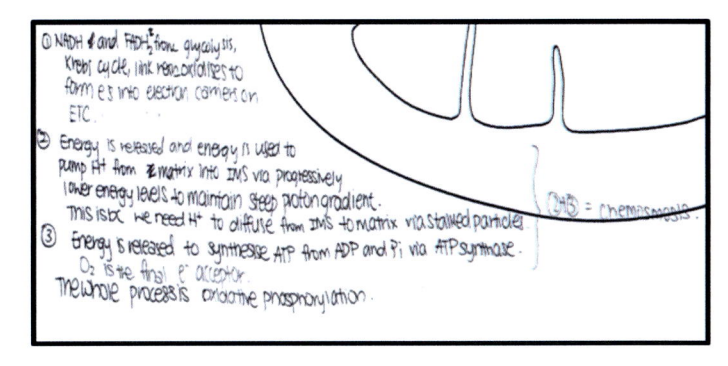

Figure 4. Student's written annotation to describe scientific process

For example, there was no further probing of the student's written text in the annotation (Figure 4) based on the earlier concept sketch which could have unpacked the precise description to convey with greater accuracy what was conveyed. Gaps in reasoning were evident as electrons were already inside these molecules (NADH and FADH2) and were therefore not formed but simply 'donated', and there was a need to specify the link reaction given the sequential process as in the following:

Student's annotation	Proposed refinement (italicised)
1. NADH and FADH2 from glycolysis, Kreb's cycle and link reaction oxidizes to form electrons into electron carriers on ETC (Electron Transport Chain.	1. NADH and FADH2 from glycolysis, *link reaction and Kreb's cycle*, oxidizes to *donate* electrons to the electron carriers on ETC (Electron Transport Chain).

In the next annotation, for greater clarity to explain where the energy came from, teacher could have guided student's attention to the electron movement along the Electron Transport Chain (ETC) and the electron progressive movement:

2. Energy is released and energy is used to pump H+ from matrix into IMS via progressively lower energy levels to maintain steep proton gradient. This is because we need H+ to diffuse from IMS (InterMembrane Space) to matrix via stalked particles.	2. *During the movement of electron along the electron transport* chain, energy is ~~released~~ *generated* and energy is used to pump H+ from matrix into IMS ~~via~~ (*as the electron moves* progressively *to the* lower energy levels *and O2 is the final electron acceptor in this chain*) to maintain steep proton gradient. This is because we *eventually* need H+ to diffuse from IMS (InterMembrane Space) *back* to matrix via stalked particles.

The precision required to explain or relate the phenomenon in terms of the purpose served 'to maintain steep proton gradient' (describing the teleological nature of processes distinctive in biology, Tamir, 1985) as opposed to explaining the how as in the causal-effect links (Yip, 2009) where the electron 'moves progressively to the lower energy levels' (mechanistic) demands an understanding of the intricacies required in the nature of biological explanations (Tamir, 1985; Yip, 2009) which could have been surfaced through the teacher's dialoguing with students.

The limitations in the student's annotation where specificity in language use to describe 'energy generated' as opposed to 'released', and description of the proton (H+) movement could have been surfaced more clearly by the teacher in discussion with students:

3. Energy is released and to synthesize ATP from ADP and Pi via ATP synthase. O2 is the final electron acceptor The whole process is oxidative phosphorylation.	3. Energy is ~~released~~ *generated* ~~and~~ *as the H+ moves through the ATP synthase complex* to synthesize ATP from ADP and Pi via ATP synthase. ** ~~O2 is the final electron acceptor~~ The whole process is oxidative phosphorylation.

8. Implications

The study shows how meaning-making through students' annotated concept sketch can be supported by teachers' use of classroom talk focused on the relevant specifics in what students have generated. The following sections elaborate on the specific areas for consideration pertaining to the place of classroom talk and specific teacher roles in relation to mediating the visual and textual modes of communication to enhance content learning.

8.1 Beyond the visual and the textual- the place of classroom talk

In this study, the student-generated concept sketch provided the initial starting point for learning. The concept sketch as a representation of students' knowledge, even if partial or inaccurate, essentially laid the ground on which the other modes of communication revolved around it and were developed. By itself, the students' visual representation would not be able to achieve as much. Teachers' use of classroom talk in the way information was elicited, questions were reformulated to direct attention, students' learning was scaffolded to draw out understanding of rationale and responses revoiced to sharpen focus was aimed at 'creating difference' (Kress et al., 2001, p. 51) through the verbal mode. Purposeful classroom talk offers, according to Kress et al. (2001), the 'overlay of detail' (p. 51) and makes transparent, through oral communication, what is relevant and necessary that is mapped onto the visual representation. Effective teacher questioning with opportunities for 'responsive questioning and feedback' (Chin, 2007) through eliciting, probing and challenging students' thinking through the pedagogical interactions (Scott, 1997) can make explicit the crucial links or connections in what was represented visually with the written annotation.

8.2 Implications for teachers' professional learning

The need for teachers to develop a culture where students' voice is valued and contributions expressed, visually or in text, through the deliberate planning of time and space for active student engagement is critical. Students can be guided to clarify ideas and build on their peers' contributions as they refine their own understanding, and sharpen how they describe and explain scientific phenomena. The study highlighted the importance of equipping teachers with the skills to orchestrate classroom discussion and foster students' engagement in their learning to reinforce the learning drawing on visual representations. This echoes researchers' identification of the need for content area teachers to provide interactional scaffolding to help students use the appropriate academic language (Lo & Macaro, 2015). Teachers can learn how to modify their questioning with the appropriate prompts to support students' content learning. Specifically, this study reinforced the need for teachers, in navigating the 'complex interweaving between the visual, linguistic' (Kress et al., 2001, p. 58) resources drawn on in communication to effectively strategize their classroom talk in specific ways:

i. revoicing to draw and focus attention in concrete terms:
talk with an indexical function that refocuses and highlights specifics to make explicit directionality, location specifics relevant to the given phenomenon in

the material context for students to re-structure thinking, re-organize specifics and/or revise representations
ii. foregrounding to make salient specific details, key features prominent:
talk that scaffolds student's thinking and challenges assumptions to help them see the key aspects, underlying rationale and reinforce links or connections across related concepts
iii. gap-filling through guiding and shaping thinking to what is relevant:
talk that mediates learning through engaging students to wrestle with content specifics and negotiate meaning-making as teachers and students 'inter-think' (Littleton & Mercer, 2013) through the exchange of ideas, generate knowledge and understanding by working with information (Mercer, 1995) to surface 'alternative conceptions' (Driver & Easley, 1978; Gilbert & Watts, 1983) and address the abstraction gaps;
and
iv. transformative in modifying perspectives, ways of seeing and thinking; sharpening specificity in language use:
talk that makes explicit the language-specific demands with attentiveness to the precision and accuracy of content specifics makes a difference to the extent to which understanding of the targeted scientific phenomenon has been represented and communicated effectively.

The study emphasised how the teacher's questioning sequences play a role in helping students to gain specificity in the language of science, particularly when it comes to reinforcing the inter-connectedness of concepts which are required in order to make sense within the larger frame of the scientific phenomenon. Given that scientific systems, processes and functions are in themselves complex in biology, this calls for the need to make the appropriate connections and relationships explicit through the appropriate use of language in order for coherence to be tightened in scientific explanations. In biology in particular, understanding the nature of teleological explanation which considers the end result as a reason for certain biological processes or structures (Yip, 2009, p.149) is valuable in enabling students to avoid misconceptions peculiar to the learning of biology (Tamir, 1985). Teachers further need to be mindful that it is insufficient to over-emphasize one concept to the exclusion of others which may be critical in understanding how concepts are linked to each other and for specificity in the accurate use of language without over-simplifying any aspects of the targeted phenomenon. Indeed, CLIL in the context of biology is not limited to learning content merely in English as a language in the content classroom. More, the nature of CLIL poses an additional challenge for students in understanding the complexities inherent in the nature of the 'languaging' of biology. In a sense, the language of biology, particularly with regard to the nature of specificity that this paper highlights in the articulation of scientific phenomena is no one's first language, but, in effect, everybody's second or possibly even third language.

The study also highlighted the extent to which teachers were able to effectively use classroom talk to draw students' attention to the gaps in learning evident in students' concept sketch. Capacity building of teachers could include guiding teachers to focus on not just the visual representations but also the written annotations provided which may surface inaccuracies as seen earlier. At the same time, the limitations of concept sketch in themselves are acknowledged in that the sequencing of processes as given in the annotations may not always be clear. This could be mitigated through numbering the annotations as seen in Figure 4 to indicate the sequence for making sense of annotations related to the concept sketch.

9. Conclusion

This study contributes to the field by exploring how teachers' classroom talk drawing on students' concept sketch could sharpen the focus on specific aspects of science concepts as students engage in meaningful dialogue with the teacher and each other. The findings demonstrate the potential of classroom talk as a 'bridging strategy' (Lin, 2016) to make explicit the relevant information involved in meaning-making using specific modalities. The study reinforces the importance of the teacher's role in facilitating targeted classroom discussion that elicits, reinforces, probes, challenges and extends students' responses to guide and sharpen students' thinking based on what is represented visually and in text. Active student engagement through carefully structured talk is aimed at scaffolding students' content learning.

The study reinforces how teacher can highlight and address students' gaps in learning through carefully structured talk which enables students to align the visual with the textual in the given context. The focus on purposeful talk that is intentional with a deictic function helps students navigate the complex nature of meaning-making across specific modalities to enhance students' content learning. This would work towards teachers helping learners, in what Moje (2018, p. vi) believes is critical, that is, to 'see in different ways' and fill the gap in grasping what they would otherwise not be able to. The findings also reiterate the need for classroom talk to make explicit the language-specific demands based on students' visual representations to sharpen the precision and accuracy of content vocabulary in specific contexts. At the same time, this study has shown the limitations in teachers' classroom talk, particularly where a focus on the visual representations to the extent of overlooking the written descriptions provided in the annotations falls short of drawing out the limitations and nuances of different aspects of concept sketches.

This paper does not claim to be exhaustive in its investigation of Biology teachers' use of classroom talk to mediate learning through students' concept sketch, given the relatively limited data set involving a select group of participants in a specific learning context. More studies could be carried out in following up on specific individual teachers' talk moves with the use of students' artefacts in different contexts with students of different ability levels. This would be invaluable to determining what it is teachers can do to effectively engage students in the process of visually constructing their content learning. Studies of this nature can contribute to a deeper understanding and inform teachers' pedagogical practices and future research.

Acknowledgements

The work reported in this article is supported by the English Language Institute of Singapore (ELIS) Research Fund under research grant ERF-2015-03-YAQ for the study funded by the Ministry of Education, Singapore. The authors would like to acknowledge the contributions of Lin Wenjie, Glendon Phua and Davina Chai in this study.

References

Ainsworth, S., Prain, V., & Tytler, R. (2011). Drawing to learn in science. *Science*, 333(6046), 1096–1097. https://doi.org/10.1126/science.1204153

Alexander, R. (2008). Culture, dialogue and learning: Notes on an emerging pedagogy. In N. Mercer & S. Hodgkinson, (Eds.), *Exploring talk in school* (pp. 91–114). London: Sage.

Bell, J.C. (2014). Visual literacy skills of students in college-level Biology: Learning outcomes following digital or hand-drawing activities. *The Canadian Journal for the Scholarship of Teaching and Learning*, 5(1). https://doi.org/10.5206/cjsotl-rcacea.2014.1.6

Bennet, D. (2011). *Multimodal representation contributes to the complex development of science literacy in a college biology class*. University of Iowa Iowa Research Online. https://doi.org/10.17077/etd.dhati9dz

Bransford, J., & Schwartz, D. (1999). Rethinking transfer: A simple proposal with multiple implications. *Review of research in education*, 24, 61–100.

Chapin, S., O'Connor, C., & Anderson, N. (2013). *Classroom discussions in Math: A teacher's guide for using talk moves to support the common core and more, Grades K-6: A Multimedia Professional Learning Resource (third edition)*. Sausalito, CA: Math Solutions Publications.

Chia, B.P., Tay, H.M., Ho, C., Ho, J., & Lee, G.B. (2014). Scaffolding scientific explanation in Chemistry through language-specific support. In Lee, Y.-J., Lim, N.T.-L., Tan, K.S., Chu, H.E., Lim, P.Y., Lim, Y.H., & Tan, I. (Eds)., *Proceedings from the International Science Education Conference (ISEC) 2014* (pp. 316–353). Singapore: National Institute of Education.

Chin, C. (2007). Teacher questioning in Science classrooms: Approaches that stimulate productive thinking. *Journal of research in Science teaching*, 44(6), 815–843. https://doi.org/10.1002/tea.20171

Collins, A. M., Joseph, D., & Bielaczyc, K. (2004). Design research: Theoretical and methodological issues. *Journal of the Learning Sciences*, 13(1), 15–42. https://doi.org/10.1207/s15327809jls1301_2

Curriculum Planning and Development Division (2016). *Biology syllabus. Pre-university. Higher 1. Syllabus 8876*. Singapore: Curriculum Planning and Development Division, Ministry of Education.

Dalton-Puffer, C. (2013). A construct of cognitive discourse functions for conceptualising content-language integration in CLIL and multilingual education. *European Journal of Applied Linguistics*, 1(2), 216–253. https://doi.org/10.1515/eujal-2013-0011

Dawes, L. (2004). Talk and learning in classroom science. *International Journal of Science Education*, 26(6), 677–695. https://doi.org/10.1080/0950069032000097424

Driver, R., & Easley, J. (1978). Pupils and paradigms: A review of literature related to concept development in adolescent science students. *Studies in Science Education*, 5, 61–84. https://doi.org/10.1080/03057267808559857

English Language Institute of Singapore (ELIS). (2011). *Whole school approach to effective communication in English*. Retrieved from <http://www.elis.moe.edu.sg/professional-learning/subject-literacy>

Esiobu, G. O., & Soyibo, K. (1995). Effects of concept and vee mappings under three learning modes on students' cognitive achievement in ecology and genetics. *Journal of Research in Science Teaching*, 32(9), 971–995. https://doi.org/10.1002/tea.3660320908

Ford, M., & Forman, E. A. (2006). Refining disciplinary learning in classroom contexts. *Review of Research in Education*, 30, 1–33. https://doi.org/10.3102/0091732X030001001

Gilbert, J. K., & Watts, D. M. (1983). Concepts, misconceptions and alternative conceptions: Changing perspectives in science education. *Studies in Science Education*, 10, 61–98. https://doi.org/10.1080/03057268308559905

Hogan, K., & Pressley, M. (1997). *Scaffolding student learning: Instructional approaches and issues*. Cambridge, MA: Brookline Books.

Jaipal, K. (2009). Meaning making through multiple modalities in a biology classroom: A multimodal semiotics discourse analysis. *Science Education*, 94(1), 48–72. https://doi.org/10.1002/sce.20359

Jewitt, C., Kress, G., Ogborn, J., & Tsatsarelis, C. (2001). Exploring learning through visual, actional and linguistic communication: The multimodal environment of a science classroom. *Educational Review*, 53(1), 5–18. https://doi.org/10.1080/00131910123753

Johnson, J. K., & Reynolds, S. J. (2005). Concept sketches – Using student- and instructor-generated, annotated sketches for learning, teaching, and assessment in Geology courses: *Journal of Geoscience Education*, 53(1), 85–95. https://doi.org/10.5408/1089-9995-53.1.85

Kawalkar, A., & Vijapurkar, J. (2013). Scaffolding science talk: The role of teachers' questions in the inquiry classroom. *International Journal of Science Education*, 35(12), 2004–2027. https://doi.org/10.1080/09500693.2011.604684

Kress, G. (2010). *Multimodality: A social semiotic approach to contemporary communication*. London: Routledge.

Kress, G., Jewitt, C., Ogborn, J., & Tsatsarelis, C. (2001). *Multimodal teaching and learning: The rhetorics of the science classroom*. London: Continuum.

Kress, G., & van Leeuwen, T. (2001). *Multimodal discourse: The modes and media of contemporary communication.* London: Arnold.

Lemke, J. L. (1990). *Talking science: Language, learning and values.* Norwood, NJ: Ablex.

Lemke, J. L. (1998). Multiplying meaning: Visual and verbal semiotics in scientific text. In J. Martin & R. Veel (Eds.), *Reading science* (pp. 87–113). London: Routledge.

Lemke, J. L. (1998). Multiplying meaning: Visual and verbal semiotics in scientific text. In J. R. Martin & R. Veel (Eds.), *Reading science: Critical and functional perspectives on discourses of science* (pp. 87–113). London: Routledge.

Lin, A. M. Y. (2016). *Language across the curriculum and CLIL in English as an additional language (ELAL) contexts: Theory and practice.* Singapore: Springer Science+Business Media Singapore. https://doi.org/10.1007/978-981-10-1802-2

Liu, Y. (2018). Literacy challenges in chemistry: A multimodal analysis of symbolic formulas. In K. S. Tang & K. Danielsson (Eds.), *Global developments in literacy research for science education* (pp. 205–218). Cham: Springer. https://doi.org/10.1007/978-3-319-69197-8_13

Littleton, K., & Mercer, N. (2013). *Interthinking: Putting talk to work.* Abingdon: Routledge. https://doi.org/10.4324/9780203809433

Lo, Y. Y., & Macaro, E. (2015). Getting used to content and language integrated learning: What can classroom interaction reveal? *The Language Learning Journal, 43*(3), 239–255. https://doi.org/10.1080/09571736.2015.1053281

Mercer, N. (1995). *The guided construction of knowledge: Talk amongst teachers and learners.* Clevedon: Multilingual Matters.

Michaels, S., & O'Connor, C. (2012). *Talk science primer.* Cambridge, MA: Technical Education Research Centers (TERC).

Michell, M., & Sharpe, T. (2005). Collective instructional scaffolding in English as a second language classrooms. *Prospect, 20*(1), 31–58.

Ministry Of Education (MOE) & University of Cambridge Local Examinations Syndicate (UCLES). (2015). *Biology Higher 2 (2017) (Syllabus 9744).* Singapore: Singapore Examinations and Assessment Board, MOE and Cambridge international Examinations.

Moje, E. (2018). Foreword. In K. S. Tang & K. Danielsson (Eds.), *Global developments in literacy research for science education* (pp. v–vii). Cham: Springer.

Novak, J. D. (1998). *Learning, creating, and using knowledge: Concept maps as facilitative tools in schools and corporations.* Mahwah, NJ: Lawrence Erlbaum Associates. https://doi.org/10.4324/9781410601629

Novak, J. D., & Wandersee, J. (Eds.) (1991). Concept mapping [Special issue] *Journal of Research in Science Teaching, 28*(10).

O'Donnell, A., Dansereau, D., & Hall, R. H. (2002). Knowledge maps as scaffolds for cognitive processing. *Educational Psychology Review, 14*, 71–86. https://doi.org/10.1023/A:1013132527007

Prain, V., & Hand, B. (2016). Learning science through learning to use its languages. In Hand, B. & McDermott, M. (Eds.), *Using multimodal representations to support learning in the Science classroom* (pp. 1–11). Cham: Springer. https://doi.org/10.1007/978-3-319-16450-2_1

Prinou, L., Halkia, K. (2003). Images of cell division on the Internet. In Constantinou & Zacharia (Eds.), *Computer based learning in science, New technologies and their applications in education* (pp. 1103–1113). Nicosia: University of Cyprus.

Reynolds, S. R., & Tewksbury, B. (2005). *On the cutting edge. Exploring teaching strategies: Concept sketch.* Retrieved from <https://serc.carleton.edu/NAGTWorkshops/coursedesign/tutorial/strategies.html>

Roam, D. (2008). *Back of the Napkin: Solving problems and selling ideas with pictures.* New York, NY: Penguin.

Roth, W.-M. (2005). *Talking science: Language and learning in science classrooms.* Lanham, MD: Rowman & Littlefield.

Schwarz, C. V., Reiser, B. J., Davis, E. A., Kenyon, L., Achér, A., Fortus, D., Scwartz, Y., Hug, B., & Krajcik, J. (2009). Developing a learning progression for scientific modeling: Making scientific modeling accessible and meaningful for learners. *Journal of Research in Science Teaching, 46,* 632–654. https://doi.org/10.1002/tea.20311

Scott, P. (1997). Developing science concepts in secondary classrooms: An analysis of pedagogical interactions from a Vygotskian perspective (Unpublished doctoral dissertation). University of Leeds.

Scott, P. (1998). Teacher talk and meaning making in Science classrooms: A Vygotksyan analysis and review. *Studies in Science Education, 32*(1), 45–80. https://doi.org/10.1080/03057269808560127

Singapore Department of Statistics (2015). *General Household Survey 2015.* Retrieved from <https://www.singstat.gov.sg/docs/default-source/default-document-library/publications/publications_and_papers/GHS/ghs2015/ghs2015.pdf>

Swain, M. & Lapkin, S. (2013). A Vygotskian sociocultural perspective on immersion education: The L1/L2 debate. *Journal of Immersion and Content-Based Language Education, 1*(1),101–129. https://doi.org/10.1075/jicb.1.1.05swa

Tamir, P. (1985). Causality and teleology in high school biology. *Research in Science and Technological Education, 3,* 19–28. https://doi.org/10.1080/0263514850030103

Tang, K. S. (2016). The interplay of representations and patterns of classroom discourse in science teaching sequences. *International Journal of Science Education, 38*(13), 2069–2095. https://doi.org/10.1080/09500693.2016.1218568

Temple, S. (1994). Thought made visible – the value of sketching. *Co-Design Journal, 1,* 16–25.

The Straits Times. (1983). It's English for all. Alfred, H. & Tan, J. *The Straits Times,* p.1. Retrieved from NewspaperSG.

Tytler, R., & Hubber, P. (2016). Constructing representations to learn Science. In B. Hand & M. McDermott (Eds.), *Using multimodal representations to support learning in the Science classroom* (pp. 159–181). Cham: Springer. https://doi.org/10.1007/978-3-319-16450-2_9

Vygotsky, L. S. (1978). *Mind in society: The development of higher psychological processes.* Cambridge, MA: Harvard University Press.

Vygotsky, L. S. (1986). *Thought and language.* Cambridge, MA: The MIT Press.

Walsh, S. (2013). *Classroom discourse and teacher development.* Edinburgh: Edinburgh University Press.

Wise, N. (2006). Making visible. *Isis, 97*(1), 75–82. https://doi.org/10.1086/501101

Yip, C.-W. (2009). Causal and teleological explanations in biology. *Journal of Biological Education, 43*(4), 149–151. https://doi.org/10.1080/00219266.2009.9656174

Zwiers, J., & Crawford, M. (2011). *Academic conversations: Classroom talk that fosters critical thinking and content understandings.* Portland, ME: Stenhouse.

Appendix A. Sample extract of lesson plan

Duration/ min	Objective	Activity
10		1. Distribute worksheet 1 to students

10	Start questionmg about stalked particle because this is the most promment feature of the mitochondria and it allows teachers to ask Why quesbons (eg Focus Area (FA) 3 – Probe for reasoning/ evidence; Challenge students' statement/assumption). This helps to scaffold students' understanding of why things work in a certain way. Scaffold quesbons step-by-step.	**Questions**
		1. Label the drfferent regions of the mitochondria:
		   a. Outer membrane
		   b. Inner membrane
		   c. Matrix
		   d. Intermembrane space
		[Teacher quickly scan through students' worksheet, check that all answers are correct. Teacher to reveal answers.]
		2. Draw the stalked partcle on the diagram
		[Teacher to walk around and identify a student who drew correctly and a student who drew wrongly. Teacher to invite both students to draw on write-board / show live scripts. Teacher to ask students to explain the orientation of stalked particles]
		[**FA3: Probe for reasoning e.g. Why is the stalked particle drawn this way?**]
		[**FA1: Revoice for verification** – *So you're saying that…/I wonder whether you mean*]

Appendix B.   Extracts of sample focus areas and talk moves

Focus area : Voicing and clarifying students' ideas	
Talk Move	**Frames for prompting**
Seek Clarification	*Can you elaborate on X?*
	What do you mean by X?
	Can you be more specific about X?

Focus area : Deepening individual students' reasoning	
Talk Move	**Frames for prompting**
Probe for reasoning or evidence	*Why do you think that?*
	How did you come up with that answer/solution?
	What's your evidence for that?

Focus area : Engaging with each other's reasoning	
Talk Move	**Frames for prompting**
Elicit students' views on other students' ideas	*What do you think about what X has just said?*
	Who would like to respond to X's idea and tell us why you agree or disagree?
	Who has a similar/different idea about how this works?

(Adapted by English Language Institute of Singapore, 2016 from Michaels & O'Connor, 2012 and Zwiers & Crawford, 2011)

Co-developing science literacy and foreign language literacy through "Concept + Language Mapping"

Peichang He and Angel M.Y. Lin
The University of Hong Kong Simon Fraser University

Drawing on Lemke's (1990) "thematic patterns" theory, this research proposes a "Concept + Language Mapping" (CLM) approach and tried it out in an English Medium Instruction (EMI) biology classroom in Hong Kong. Lessons were observed and samples of student work were collected during the intervention with student/teacher interviews conducted afterwards. A quasi-experimental design was also adopted to estimate the impact of the CLM approach. The analysis indicated that CLM facilitated the development of both content and language knowledge.

Keywords: "Concept + Language Mapping" (CLM), Content and Language Integrated Learning (CLIL), English medium instruction (EMI), foreign language literacy, science literacy, thematic patterns

1. Introduction

Content and Language Integrated Learning (CLIL) is an educational approach where students learn non-language content subjects through a second/foreign/additional language (L2) (Coyle, Hood, & Marsh, 2010) and it has become a key research domain in bilingual/foreign language education. Previous literature on CLIL largely focused on its various definitions, language and content learning outcomes and pedagogical issues (Cenoz, Genesee, & Gorter, 2014; Dalton-Puffer & Nikula, 2014; Lin, 2016; Llinares, Morton, & Whittaker, 2012). Recent studies have begun to explore CLIL teacher education (Cammarata & Ó Ceallaigh, 2018) such as CLIL teachers' knowledge about language (Morton, 2018), professional identities (Dale, Ron, & Verspoor, 2018) and language awareness (He & Lin, 2018). While there is no denying that CLIL involves the teaching of both content and language, it remains a challenge to achieve pedagogical integration of content and language in CLIL classrooms (Dalton-Puffer, 2018; Lin, 2016; Ruiz de Zarobe,

2016). Although researchers have explored the balance between language and content (Cammarata & Tedick, 2012; Dalton-Puffer, 2013; Lyster, 2007) and conceptualized the integration of the two (Nikula, Dafouz, Moore, & Smit, 2016), these studies have mainly drawn on language pedagogy rather than content pedagogy perspectives (Cenoz, 2016; Dalton-Puffer, 2018). A "notable gap" has been "the lack of involvement" of subject specialists in CLIL research (Dalton-Puffer & Nikula, 2014, p. 119) incorporating "expert perspectives of subject education researchers" (Dalton-Puffer, 2018, p. 386). Echoing the need for both language and subject-specific perspectives on CLIL, Lin (2016) recommended using "thematic patterns", a notion proposed by Lemke (1990) in his seminal work *Talking Science*, to integrate content and language pedagogies across the curriculum.

Science academic literacy is both cognitively and linguistically demanding; hence, learning science is virtually learning a "foreign language" (Wellington & Osborne, 2001). In science classrooms where content subjects are taught in English as an additional language (EAL), the learning task is "foreign language squared" (i.e. two "foreign languages" multiplying each other: a cognitively unfamiliar "foreign language" X a linguistically "foreign language" = a learning task of foreign language^2). In this study, we address issues concerning the co-development of science literacy and academic language literacy in English medium instruction (EMI) science education. Drawing on Lemke's (1990) theory of thematic patterns, we proposed the "Concept + Language Mapping" (CLM) approach as an innovative CLIL pedagogy and pioneered it in an EMI biology class in a secondary school in Hong Kong. The impact of CLM pedagogy is examined and the findings and implications for CLIL teacher education will be discussed.

2. Literature review

This section reviews the research traditions of concept mapping and thematic patterns respectively. Both concepts contribute to the theoretical framework for the CLM approach.

2.1 Meaningful learning and concept mapping

Education of all content subjects involves learning of concepts which are traditionally defined as "a perceived regularity in events or objects designated by an arbitrary label" (Novak, Gowin, & Johansen, 1983, p. 625). Concepts are seen as abstract but are fundamental for all content subjects, and they remain challenging for classroom teaching. Grounded in the assimilation theory of meaningful learning (Ausubel, 1968), Novak et al. (1983) contributed to concept learning by developing the strategy of concept mapping, which has been widely applied in the teaching of various subjects. Concept mapping is believed to facilitate meaningful learning by constructing a spatial and visual representation of interconnected concepts and the hierarchical structure of conceptual knowledge in the human mind (Novak et al., 1983; Novak, 2010).

2.2 Thematic patterns

Although concept mapping has been regarded as a useful meta-cognitive strategy, it has limitations arising from its neglect of the role of language in learning. As pointed out by Novak (2010), concept maps which "strip away all text except for concept labels" may lead to the "lack of clarity for most people" (p. 32). It is also noted that although concepts in concept maps are linked in a meaningful way to indicate the interrelationship between concepts, the concepts themselves may be too abstract for learners to understand. Such mentalistic representations of concepts, according to Lemke (1998), "lacks the necessary vocabulary" to tell teachers what they should do to help students to understand the concepts.

Rather than conceptualizing "concept" as abstract mental representations of objects or events, Lemke (1990) proposed the notion of "thematic pattern", defined as "the pattern of connections among the meanings of words in a particular field of science" (Lemke, 1990, p. 12). According to Lemke, each specialized field of human activity has its own unique semantic patterns (i.e., conceptual system). Within each thematic pattern, there are "thematic items" linked by their customary semantic relationships. On the one hand, each thematic pattern can be "condensed" and become a thematic item of another thematic pattern at a higher semantic hierarchy; on the other hand, thematic patterns in different parts of the specialized field can be interrelated to form a more complex "thematic nexus" (i.e., a synthesis). In order to communicate ideas, we need to express relationships between the meanings of different thematic items. Language is a system of resources for making meaning and it is used to describe not only the semantic relationships between different thematic patterns but also those within a particular thematic pattern. For example, the concept "photosynthesis" (Figure 1) can be conceptualized as a thematic pattern consisting of "thematic items" (i.e., process, green plants, food, carbon dioxide, water, and light energy) connected by different specific "semantic relations" (i.e., TOKEN/TYPE; AGENT/PROCESS; PROCESS/TARGET; and CIRCUMSTANCE: manner/material/condition). The thematic pattern "photosynthesis" can be "condensed" to a thematic item and woven into another thematic pattern expressing the "reason for the importance of photosynthesis" (i.e., the sentence "Photosynthesis is important because it produces food (starch) and releases oxygen for all living things"). The latter thematic pattern may in turn become one of the many logically related themes (thematic units) and be further woven into a text (i.e., "thematic nexus") about "how green plants obtain energy" in a science lesson.

Photosynthesis (   ) is the process by which green plants make food from carbon dioxide and water using light energy.

The semantic relations in the definition "photosynthesis"
1. PHOTOSYNTHESIS is a PROCESS [Token / Type]

2. GREEN PLANTS make FOOD [Agent / Process / Target]
–by PHOTOSYNTHESIS [Circumstance: manner]
–from CARBON DIOXIDE and WATER [Circumstance: material]
–using LIGHT ENERGY [Circumstance: condition]

An example of a thematic pattern

Photosynthesis is important
            [Carrier / Attribute]

because
[logical relation: Cause/ Consequence]

it produces food (starch)   and   releases oxygen   for all living things.
[Agent / Process / target] [logical relation: Item/ Addition] [Process/target]   [Circumstance: beneficiary]

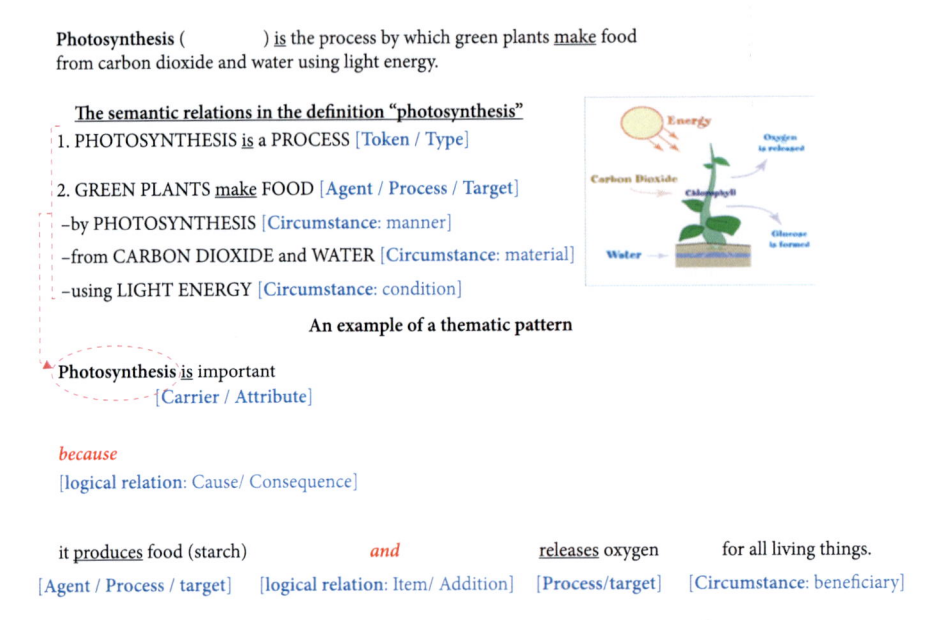

Figure 1. Thematic patterns and semantic relations about photosynthesis

In this way, Lemke's (1990) thematic pattern theory offers teachers a linguistic tool which enables them not only to de-construct/analyze the thematic items and semantic relations within/between different concepts (i.e., thematic patterns) but also, through "theme-weaving", to establish thematic interconnections between different thematic patterns at more than one intermediate thematic nexus (in traditional terms: linking concepts learned in different lessons).

2.3 Thematic-pattern-based "concept + language mapping"

Drawing on Lemke's (1990) thematic pattern theory and previous research findings, the present study extended "concept mapping" (Novak et al., 1983) to a thematic-pattern-based CLM approach (Figure 2) by emphasizing the role of language in concept instruction in CLIL lessons.

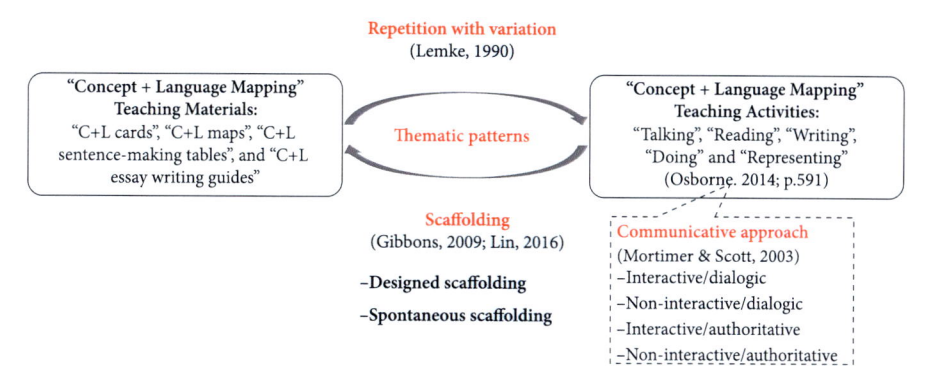

Figure 2. The thematic-pattern-based CLM approach

In the CLM pedagogy, CLM materials[1] including C+L cards, C+L maps, sentence-making tables and essay writing guides are designed to present the key thematic patterns under a thematic topic. Using these materials and activities, students are enabled to engage with the thematic patterns multiple times through "repetition with variation" (Lemke, 1990). For example, the same thematic patterns in the C+L map "The process of photosynthesis" are presented in the textbook with both verbal texts and graphic diagrams; during CLIL lessons they are introduced by the teacher and discussed among students, and then are read by students in the task sheets. The thematic patterns are further explored in experiments conducted by the students and are written out in their assignments and tests. Through a series of communicative activities – "re-presenting" (Lemke, 1998), "talking", "reading", "doing" and "writing" (Osborne, 2014, p. 591), the thematic patterns focusing on the same thematic topic appear time and again, "with some items and relations similarly expressed and others differently expressed" (Lemke, 1990, p. 227). Such repetition of thematic patterns with variation helps students to understand the abstract conceptual patterns on the one hand and consolidate both content and language knowledge on the other. Both the CLM materials and CLM activities are two core components in the thematic-pattern-based CLM pedagogy. Effective implementation of the pedagogy relies on the scaffolding provided by the teacher who is the designer and instructor of the lessons. The CLM materials are part of the designed scaffolding (Gibbons, 2009; Lin, 2016) prepared by the teacher before the lessons, but they take effect only when they are flexibly activated and used by students themselves during the CLM activities; in other words, the thematic-pattern-based teaching materials must be fully understood (rather than rote-memorized) and flexibly employed during argu-

---

1. "C+L" is the abbreviation for "Concept + Language", e.g., "C+L cards" means "Concept + Language cards". Examples of the materials are shown in the results and analysis section.

mentation and inquiry of content knowledge. The role of the CLIL teacher is most significantly reflected in his/her flexible application of both designed scaffolding and spontaneous scaffolding (Gibbons, 2009; Lin, 2016) during which the teacher "talks" about content and language knowledge with students through a "communicative approach" that shifts between different combinations of interactive/non-interactive and dialogic/authoritative styles (Mortimer & Scott, 2003) according to different purposes at different stages of CLIL teaching.

3. Methodology

To gauge the potential impact of the thematic-pattern-based CLM pedagogy as well as its feasibility in CLIL classrooms, we developed the following research questions:

1. Does the CLM approach facilitate development of both content knowledge and language knowledge in the EMI biology classroom?
2. What are the processes involved in pioneering the CLM approach in an EMI biology classroom?

3.1 Research design

English remains the socioeconomically dominant language and the most important medium of instruction in Hong Kong even after its handover to China in 1997. Due to high parental pressure for EMI education, many schools offer EMI classes even though there is not enough support provided for students learning content in English (Lin & Man, 2009). This research was conducted in a Secondary 4 (S4) EMI biology class to pioneer the CLM approach to support students' learning of both content and language. Following Reeves' (2000) "development research" framework, the study proceeded in four phases: First, the researchers had pre-intervention interviews with teachers probing what they perceived as teaching and learning challenges; second, the thematic-pattern-based CLM materials (i.e., C+L cards, C+L maps, C+L sentence-making tables, and C+L essay writing guides)[2] were designed by the researchers and reviewed by

---

2. The technical terms such as "Agent", "Process", "Target", and "Circumstance" in the analysis of the CLM material examples in Figures 1, 4 and 6 are metalinguistic analytical constructs based on Lemke's Semantic Relations for Thematic Analysis (Lemke, 1990, p. 221–224). They did not appear in the CLM materials and were not taught to the students, but are indicated in the examples of this article to illustrate the semantic relations in the thematic patterns of the CLM materials.

teachers in different content areas; third, the CLM materials were tried out in lessons by the participating teachers who decided on which materials to use and when to use them; and fourth, the teachers and researchers co-reflected on the CLM pedagogy and improved it in an ongoing process. In the senior biology subject, as the research adopted a quasi-experimental design but there was only one biology class in each grade, the S4 and S5 biology teachers adjusted their teaching scheme so that the same topic "monohybrid inheritance", originally a S5 unit, could be taught to the S4 and S5 classes during the same period. Both the S4 teacher (i.e., Miss T) and the S5 teacher were experienced science teachers qualified to teach biology in English. According to the teachers, the students in S4 and S5 were both well-motivated in learning. However, although their English language proficiency was above-average among same-grade students in the city, their academic language literacy remained insufficient which affected their learning of the content subjects (e.g., biology and geography) in English as an additional language. During the intervention, pre-test and post-test were administered for the two classes on the same day. The number of biology lessons and the length of each lesson were equal for both classes following the school regular schedule, with S4 (i.e., the intervention class) adopting the CLM pedagogy while S5 (i.e., the control class) only participated in the pre/post-tests without trying out the CLM materials or activities.

Table 1. Details of control class and intervention class

	Intervention class	Control class
Grade	S4 (Grade 10)	S5 (Grade 11)
Number of students	N = 28 (gender evenly distributed)	N = 30 (gender evenly distributed)
Age	15–16 years old	16–17 years old
L1	Cantonese	Cantonese
English language proficiency	above-average among same-grade students in the city	above-average among same-grade students in the city
Discipline	well-motivated with good learning attitude	well-motivated with good learning attitude
Teacher	experienced Hong Kong local science teacher; female; Cantonese as L1 but qualified in teaching science in English; first time collaboration with the project	experienced Hong Kong local science teacher; female; Cantonese as L1 but qualified in teaching science in English; first time collaboration with the project

Table 1. *(continued)*

	Intervention class	Control class
Number of lessons during intervention	8 lessons at the same teaching weeks in the school schedule	8 lessons at the same teaching weeks in the school schedule
Pre-/post tests	taken on the same day as those for the control class	taken on the same day as those for the intervention class
Teaching resources	tried out during intervention	not available

3.2 Data collection and analysis

The concurrent triangulation mixed methods strategy (Creswell, 2003) was used to confirm, cross-validate, or corroborate the findings with both quantitative and qualitative data. A quasi-experimental design was employed to estimate the impact of CLM pedagogy in facilitating students' development of content knowledge and language knowledge. Both the pre-test and post-test examined similar content knowledge from the same unit using the same question types including multiple-choice questions, blank-filling, short questions, long questions or essay questions. To reduce testing effect, the teacher did not check answers with students after the pre-test. The quantitative data from the pre-test and post-test were collected at the beginning and end of intervention respectively. During the intervention, the researchers observed and videotaped the biology lessons taught using the CLM approach. After the intervention, a 30-minute focus group interview was conducted with five students of different academic achievement levels to probe their feedback on the CLM pedagogy; and a semi-structured interview was conducted with the teacher to probe her reflection on the intervention. The quantitative data included the pre-test and post-test scores of the control and intervention classes. The qualitative data consisted of approximately 280 minutes of videos of eight lessons, 75 minutes audio-taped interviews (a 30-minute focus group interview with students and a 45-minute semi-structured interview with Miss T), and pre/post-test papers in both the control and intervention classes. Since different types of data were collected concurrently in one research phase using both quantitative and qualitative approaches, this made it possible for the researchers to offset the weaknesses of one method with the strengths of the other. According to Creswell (2003), the concurrent triangulation mixed methods strategy produces "well-validated and substantiated" findings as it generally integrates the results of both qualitative and quantitative methods during the interpretation phase which "can either note the convergence of the findings as a way to strengthen the knowledge claims of the study or explain any lack of convergence that may result" (p. 217).

The quantitative data were analyzed following two steps: First, the pre-test and post-test were scored by a science-major research assistant and two research assistants who had graduated from a MEd CLIL program. The science-major research assistant scored the content knowledge in the tests according to the answer key checked by both the research team and the participating teachers. The CLIL research assistants scored the language knowledge of the answers in structured questions or short essays. A marking scheme for language knowledge was designed based on the CLIL practices and principles guided by genre theories of the Sydney School (Lin, 2016; Rose & Martin, 2012); namely, scores were given to correct use of language features at different levels including subject-specific vocabulary, general academic vocabulary, logical connectors, sentence patterns of academic functions, complete sentences and proper text structures. Drawing on the theoretical principle that language and content are always related (Halliday, 1993) and to avoid rote-memorized answers which were linguistically correct but irrelevant to the content topic, only correct language features in test items where the content knowledge question was also correctly answered were scored. The inter-rater correlation for the language knowledge scores was 90%. Second, the pre-test and post-test scores in the control and intervention classes were compared by independent sample *t*-tests to examine whether the CLM approach might have made a significant difference between the two classes. To minimize the confounding effects of the prior differences between the two classes, ANCOVA was performed to compare the post-test scores in the intervention class with those in the control class, using their pre-test scores as the covariate.

The qualitative analysis involved three types of qualitative data to allow for triangulation of different research findings (Creswell, 2003). First, the observed lessons were analyzed iteratively focusing on the episodes where the CLM materials were applied. The lessons were transcribed verbatim, and the transcripts were then analyzed using the conversation analytic method of sequential analysis (Lin, 2007). Second, to corroborate the researchers' interpretations of the lesson observations, the first author interviewed both the teacher and the students to explore their perceptions and feedback about the CLM pedagogy tried out in their classes. During the interview, the teacher and the students elaborated on how they used the CLM materials which helped the researchers to better understand whether the CLM approach facilitated the students' language and content development and how the teacher used the CLM resources to guide the students to better understand the biology topics. Third, the researchers' interpretation of both the lesson observation and interview data was corroborated by post-test results and the analysis of the essay question answers produced by students in both the control and intervention classes.

4. Results and analysis

This section addresses the research questions using both quantitative and qualitative data, which are analyzed based on the conceptual framework of the study.

4.1 The CLM approach facilitated content and language development in the EMI biology classroom

The first research question will be addressed by the quantitative data in Section 4.1.1 and the qualitative data in Section 4.1.2.

4.1.1 *Quantitative results*

The quantitative data included pre-test and post-test scores. Tables 2 and 3 summarized the statistics on content and language knowledge development of both the control and intervention classes.

Table 2. Summary of *t*-test results on content knowledge development

	Control class M	Control class SD	Intervention class M	Intervention class SD	*t*-test
Pre-test	11.938	2.435	10.250	3.273	0.026
Post-test	23.650	6.987	30.250	4.906	0.000

Notes. M = Mean; SD = Standard Deviation.
* $p<.05$. *** $p<.001$.

Regarding content knowledge development, Table 2 shows that the pre-test mean score of the control class ($n=32$, $M\_pre\text{-}con=11.94$, $SD=2.44$) was higher than that of the intervention class ($n=28$, $M\_pre\text{-}int=10.25$, $SD=3.27$); while in the post-test, the mean score of the control class ($n=30$, $M\_post\text{-}con=23.65$, $SD=6.99$) was lower than that of the intervention class ($n=28$, $M\_post\text{-}int=30.25$, $SD=4.91$). An independent-samples *t*-test was conducted to determine whether differences existed between the control and intervention classes in their pre-tests and post-tests. The *t*-test results indicate that the difference between the two classes in the pre-test were statistically significant, but the difference was not very large with a *p*-value slightly smaller than .05 ($t\,(58)=2.28$, $p=.026<.05$). Whereas in the post-test, the *t*-test results reveal that the difference between the two classes was highly significant with a p-value smaller than .001 (t (52) = −4.19, p = .000 < .001). After eliminating confounding effects of the pre-tests, an ANCOVA showed a strong effect of group difference: F (1, 54) = 26.08, p = .000 < .001, η_p^2 = 0.33.

Table 3. Summary of t-test results on language knowledge development

	Control class		Intervention class		
	M	SD	M	SD	t-test
Pre-test	2.781	1.237	3.107	1.771	0.408
Post-test	16.667	6.307	27.786	6.754	0.000

Notes. M = Mean; SD = Standard Deviation.
* p < .05. *** p < .001

Concerning language knowledge development, Table 3 shows that the pre-test mean score of the control class (n = 32, M_pre-con = 2.78, SD = 1.24) was lower than that of the intervention class (n = 28, M_pre-int = 3.11, SD = 1.77); while in the post-test, the mean score of the control class (n = 30, M_post-con = 16.67, SD = 6.31) was still lower than that of the intervention class (n = 28, M_post-int = 27.79, SD = 6.75). An independent samples t-test was run to decide whether differences existed between the control and intervention classes in their pre- and post-test scores. While the t-test results indicate that the difference between the two classes was not statistically significant in the pre-test, with a p-value larger than .05 (t (58) = −.83, p = .41 > .05); in the post-test, the difference between the two classes was highly significant with a p-value smaller than .001 (t (56) = −6.48, p = .000 < .001). After removing the confounding factor of pre-test, the ANCOVA result revealed a strong effect of group difference, F (1, 54) = 39.27, p = .000 < .001, η_p^2 = 0.42, with very strong observed power at 1.00. These results indicate that the intervention was highly likely to have had a positive impact on students' content and language development.

4.1.2 Qualitative results

The quantitative results were corroborated by the qualitative data. The interviews with both the teacher and students about their feedback to the CLM approach turned out to be positive. As shown in Appendix 1, the students found the CLM materials "*helpful*", "*useful*", "*beneficial*", "*good*", "*better than the textbook notes*", "*simple and clear*", "*make learning easy*", "*help me learn*", and "*help me better understand the words*"; they also found the C+L activity – a concept guessing game, "*fun and engaging*". According to the students, the reasons behind the positive comments included: first, the "*key words*" (i.e. the thematic items in the thematic patterns) in all materials were *"bold"* (subject-specific vocabulary as MEDIUM, AGENT or TARGET), *"underlined"* (the key verbs as PROCESS) and *"bold and italic"* (logical connectors) so that the students kept focusing on the thematic patterns and semantic relations in the materials; second, the C+L cards and maps summarized the *"key points"* in *"complete sentences"* introducing *"entire points"* and *"whole processes"* so that the materials *"included all relevant concepts"* and the relations among the concepts were *"simple and clear"* which allowed the students to retrieve information from a *"more focused"* knowledge domain without *"skipping the points"*; third, the sentence-making tables highlighted the main functions and logical relations (e.g., defining, expressing cause and effect) which helped the students understand the meaning of the thematic items and the interrelations within/between the thematic patterns; fourth, rather than providing only verbal notes, the CLM materials contained multimodal information (e.g., diagrams and arrows) which enabled students to *"visualize the concepts"* more easily; fifth, the CLM material-based questions-and-answers during the lessons made the content knowledge *"more impressive"*; and sixth, the CLM materials enhanced students' language knowledge including spelling, pronunciation, and the uses of every day and academic vocabulary. Students longed for more CLM materials in future lessons which they would use for distinguishing and memorizing concepts and making notes during self-directed learning.

Just like the students, Miss T also had favorable feedback on the CLM materials. She commented that the students had mastered the lesson *"quite well"* even though it was an abstract unit, and ascribed this to the CLM materials which *"included all key concepts ranging from simple to complex"*. Miss T found the C+L cards *"very clear"*, *"including all key points"* but appear *"more focused"*. She summarized three points of the C+L maps which she *"appreciated"*. First, since diagrams were presented next to the corresponding concepts, they enabled students to *"visualize"* the abstract concepts easily. Second, the interrelatedness of visual and verbal information not only helped students understand the meaning of the concept but also reminded them of its structural representation. Compared with the bullet-point summaries in textbooks, Miss T thought the C+L maps are more helpful because their multimodal features *"give students more ideas"*; more importantly, the CLM materials encouraged students to learn concepts and the interrelationship between concepts by meaningful learning rather than *"rote learning"*. A third feature of C+L maps which Miss T described as *"really good"* is the sequencing function. With the C+L maps presented via PowerPoint, the different concepts and the interrelations between concepts were not shown all at once in a huge fixed map, but appeared one by one according to the growing complexity of the concepts, which *"helped students learn the logic"* not only about how concepts in different C+L cards were linked, but also how different concept networks in different C+L maps were interconnected. Such sequencing which enabled teacher-student co-construction of concept meaning was also appreciated by students, such as S5, who said "*it really makes learning easier because the concepts can be put one by one back into the C+L Map*". As for the sentence-making tables and essay writing guides (Figure 3), Miss T was sure that both materials helped "*raise students' language awareness*". According to the teacher, due to the lack of academic writing skills, students tended to rote-learn the bullet-point notes in the textbook without paying attention to science literacy or academic language literacy. The two types of CLM materials raised students' awareness about the logical relationships in the texts, the need to elaborate on arguments, and the use of subject-specific and general academic vocabulary in academic writing.

Cause and effect			
X (result)		*because*	Y (cause)
Children look like their parents in some ways		*because*	they get their in-born characteristics from their parents.
DNA is a stable molecule		*because*	it has strong sugar-phosphate backbones and a double helix structure maintained by the hydrogen bonds between the two strands.
Since	Y (cause)	X (result)	
Since	DNA molecule has a long sequence of bases to form genetic code,	it stores a large amount of genetic information	
Y (cause)		*Therefore,*	X (result)
DNA molecule consists of a large number of nucleotides.		*Therefore,*	it carries a large amount of genetic information.
Y (cause)		*As a result.*	X (result)
DNA can replicate itself accurately through complementary base pairing.		*As a result,*	identical genetic information can be passed to the new cells from generation to generation.

Add an **introduction** referring to the question

Provide **supporting details** to make your arguments *solid*.

Use **sequencial conjunctions** to make your argument *clear*.

Use **logical connectors** (e.g. cause & effect) to make your arguments *logical*.

Use **academic words** (e.g. 'replicate' instead of 'copy') to make your arguments *scientific*.

The structure of DNA is well suited to its function as a genetic material *because of* the following aspects:

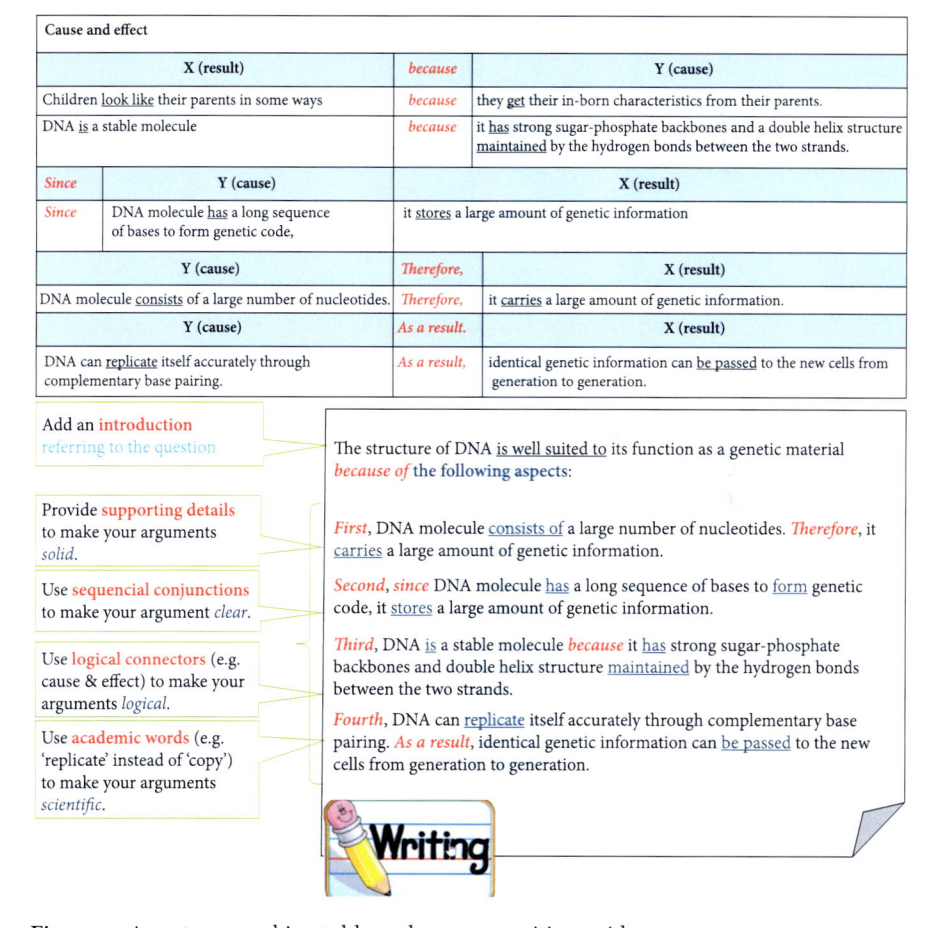

First, DNA molecule consists of a large number of nucleotides. *Therefore*, it carries a large amount of genetic information.

Second, since DNA molecule has a long sequence of bases to form genetic code, it stores a large amount of genetic information.

Third, DNA is a stable molecule *because* it has strong sugar-phosphate backbones and double helix structure maintained by the hydrogen bonds between the two strands.

Fourth, DNA can replicate itself accurately through complementary base pairing. *As a result*, identical genetic information can be passed to the new cells from generation to generation.

Figure 3. A sentence-making table and an essay writing guide

We shall further discuss these in the next section where qualitative data from class observations will be analyzed to explore how the CLM approach facilitated content and language knowledge development in the EMI biology classroom.

4.2 Integrating content and language learning with thematic-pattern-based designed and spontaneous scaffoldings in shifting communicative approaches

Before Miss T's lessons, the CLM materials were distributed to the students. They served as designed scaffolding which provided the teacher and students with shared materials to do in both self-directed and collaborative learning.

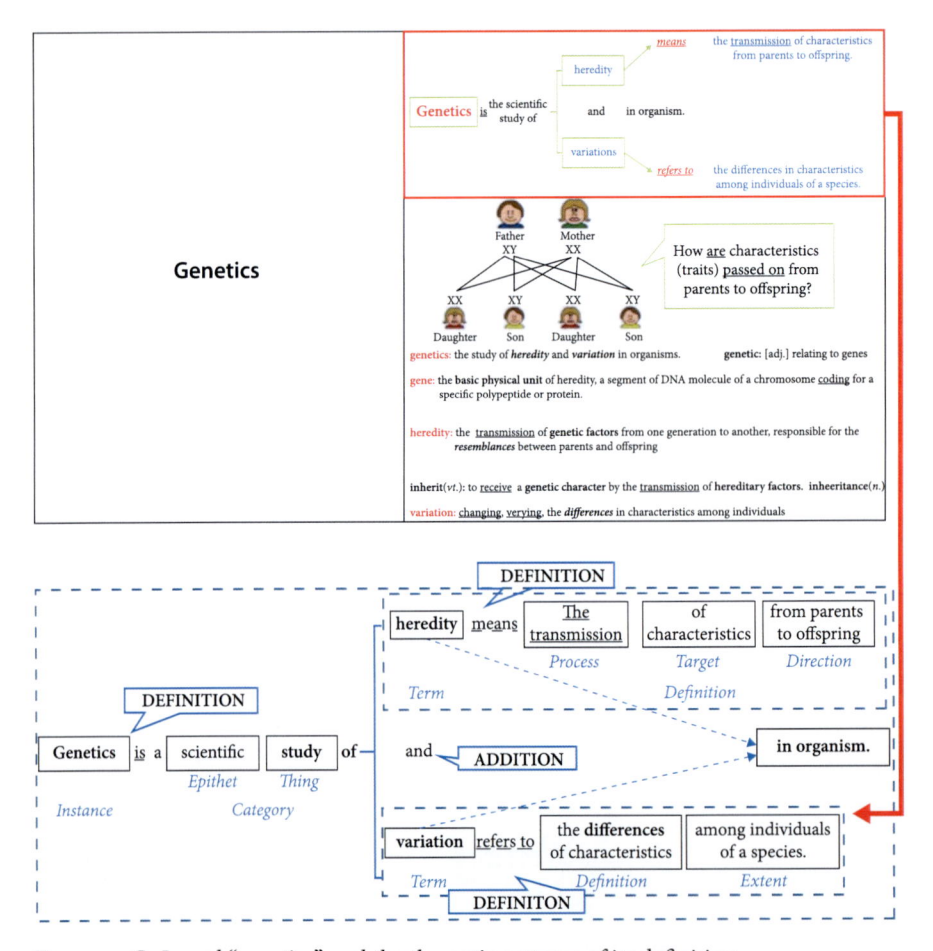

Figure 4. C+L card "genetics" and the thematic patterns of its definition

For example, when learning the concept "genetics", students were able to identify the definitions of the three interrelated concepts in the C+L card (Figure 4). Thus, instead of just noticing the definition of "genetics", students learned simultaneously the interrelationships embedded in the thematic pattern of this highly condensed concept (i.e., "genetics") which guided them to further de-construct the more detailed thematic items and interrelated semantic relations of its two basic thematic items (i.e., "heredity" and "variation") in their own corresponding thematic patterns. The C+L cards as designed scaffolding were a useful self-directed learning tool not only during the lessons but also after school.

4.2.1 *Multimodal animated sequential "concept + language mapping" with thematic-pattern-based designed and spontaneous scaffoldings*

With key concepts taught, the teacher guided students to build up interconnection between different concepts by using a C+L map which was converted to a blank-filling worksheet (Figure 5) so that students could recap concepts and discuss them with peers.

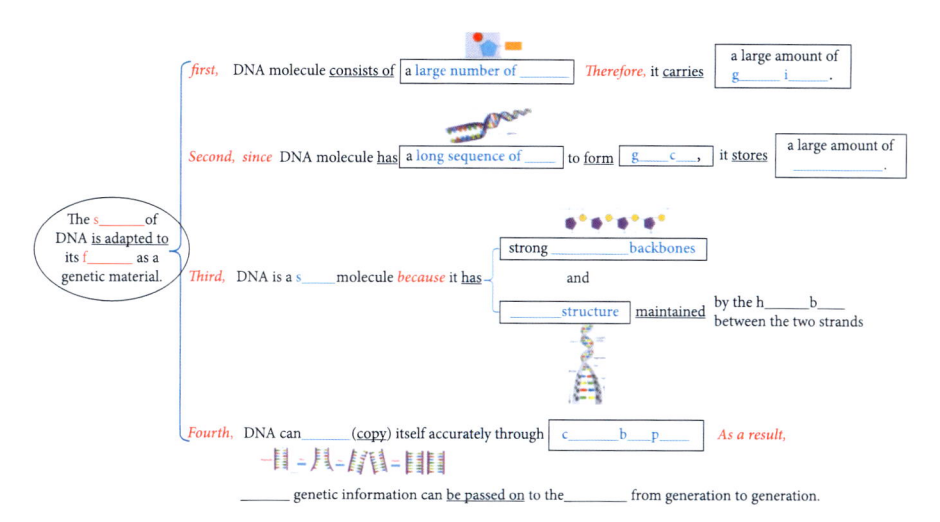

Figure 5. Blank-filling C+L map "DNA structures and function as a genetic material"

The students searched their CLM materials for answers to the C+L map worksheet. After completing the exercise by themselves, they discussed in groups. During discussion, they compared answers, questioning each other and justifying their own answers by showing evidence from the CLM materials or the textbook. According to Miss T, the peer discussion, as an "interactive/dialogic" communication approach (Mortimer & Scott, 2003) was necessary as it offered students opportunities to review the concepts, negotiate understandings, explore new ideas and correct misconceptions about the lesson. The C+L map worksheet and the other CLM materials as designed scaffolding thus became useful artefacts and references facilitating collaborative learning during class.

As the C+L map summary was composed of different themes and thematic patterns linked from different parts of the textbook, it formed a thematic nexus (Lemke, 1990) – a global network of meaning relationships with patterns of thematic items interconnected and further embedded in more complex patterns of semantic relationships. For example, Figure 6 illustrates thematic patterns and thematic items that co-construct the thematic nexus and semantic relationships within the global thematic pattern. The thematic topic "The structure of DNA is adapted to its function as a genetic material" was a thematic nexus composed of four themes, each of which consisted of a "cause and consequence" relationship with the "cause" and "consequence" both constituted by thematic patterns of interconnected thematic items. Each thematic item itself was highly condensed which could be further unpacked into thematic items and semantic relations of its own. For instance, the thematic items "nucleotide" and "complementary base paring" both have their own thematic patterns which had been shown in different C+L maps (i.e., "What is nucleotide and nucleic acids?" and "What is DNA?") elaborated in the previous lessons. With such a huge thematic web woven by complex semantic relationships as well as abstract and highly condensed thematic items, the blank-filling C+L map as a summary of the lessons could be very difficult for some of the students. The CLM materials (cards and maps) as well as the summary worksheet turned out to be useful designed scaffolding for self-directed revision and inquiry of the lesson, and the interactive/dialogic peer discussions were also beneficial collaborative learning activities.

It should be noted that the highly abstract biology lesson was not only facilitated by the CLM materials as designed scaffolding but also by the thematic-pattern-based spontaneous scaffolding (Gibbons, 2009; Lin, 2016) provided by the teacher during classroom interactions. Though by self-directed learning and peer discussion students were able to complete the C+L map worksheet, they needed to verify their thinking through further negotiation with the teacher by co-constructing the C+L map in a series of teacher-student triadic dialogues (Lemke, 1990). Such triadic dialogues were orchestrated with multimodal "C+L Mapping" facilitated by the special design of the C+L map on a PowerPoint slide. To allow the teacher and students to jointly construct the C+L map, the concepts in the map did not appear on the slide all at once, but was shown item by item according to the sequence in which the thematic patterns were discussed. This enabled both the teacher and students to focus on the same thematic pattern. The C+L map became bigger and increasingly complex as more thematic items were identified and more thematic-patterns explained, and such sequence of discus-

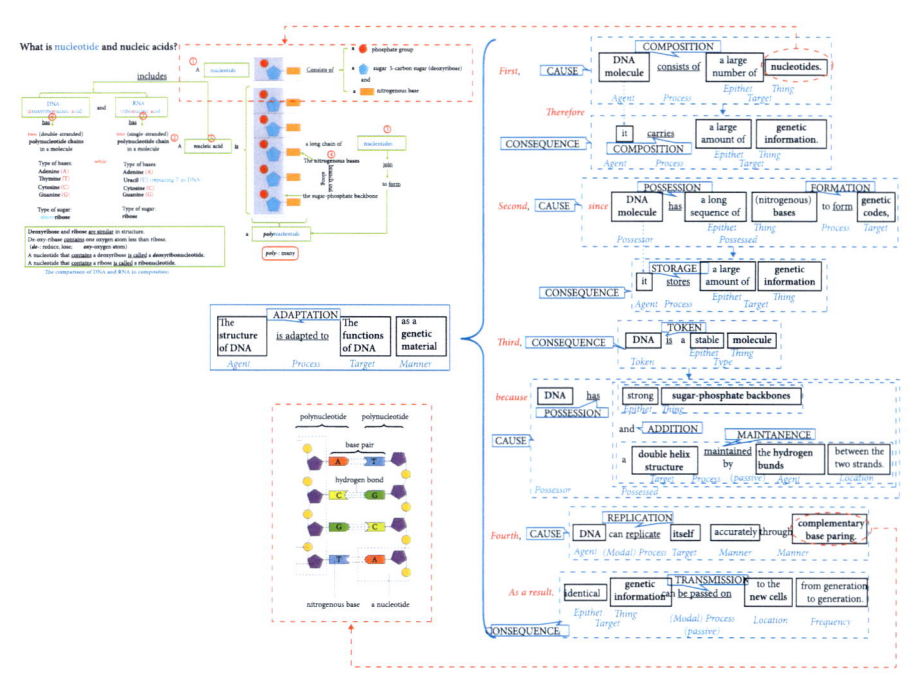

Figure 6. Thematic patterns, thematic nexus, and semantic relationships

sion about the themes, according to Miss T, fits with the strategy of *"from simple to complex"*. The questioning was based on the thematic patterns in each theme. During co-construction of the C+L map, Miss T did not just probe students for answers to the missing item of a particular thematic pattern in the blank-filling C+L map, she also asked students to define the thematic item by identifying its own condensed thematic pattern and corresponding thematic items. For example, in one scenario, after EV provided the answer *"a large number of nucleotides"*, Miss T asked students to further elaborate on the composition of nucleotide which could be prompted by the diagrams next to the concept in the C+L map. According to Miss T, the diagrams in the C+L map *"really helped students think"* because *"when students read the diagram several times, they know what concepts the diagrams represent"*. Thus, apart from asking students to recap the basic components of the structure of "nucleotide", she also reminded them to focus on the details of the corresponding diagram so that *"they also know the structure of the concepts and are able to draw them out"*.

The "interactive/authoritative" communication (Mortimer & Scott, 2003) guided by Miss T was also crucial as there were students who could not fully understand at once the thematic patterns in the complicated C+L map, hence needed spontaneous scaffolding from the teacher to help them clarify the concepts. In the following teacher-student interaction, Miss T guided MG to figure out the thematic item "nitrogenous base" in the C+L map.

Excerpt 1. Thematic-pattern-based spontaneous scaffolding in interactive/authoritative communication

T	Second. What will be the answer? MG
MG	Second, since DNA molecule has a long sequence of nucleotide…
T	Ah… DNA molecule has a long sequence of nucleotides. We know that it contains a sequence of…
MG	[correcting himself] A long sequence of genes.
T	A long sequence of genes. But what? Which part in the genes to form the…
MG	Chromosome.
T	Form the chromosome? No… The DNA molecule has a long sequence of which structure to form the …
MG	Form the genetic code.
T	Form the genetic code. Very good. But which part of the nucleotide? Which part of the nucleotide form the genetic code? This is the point. The phosphate group? Or the…
MG	Nitrogenous base.
T	Nitrogenous base. Very good. Here you will find that DNA molecule has a long sequence of …
MG	Nitrogenous base.
T	Nitrogenous bases, or a long sequence of bases to form …
MG	Form the genetic code.
T	To form genetic code.

From the guiding questions, the revision of the structure of nucleotide during the discussion of the previous theme turned out to be important. As Miss T and KG had just recapped the components about the structure of nucleotide, MG could remember the structure clearly; hence, when the teacher provided the spontaneous scaffolding as a prompt ("*But which part of the nucleotide? Which part of the nucleotide form the genetic code? This is the point. The phosphate group? Or the…*"), he could utter the answer "nitrogenous base" at once and then went on jointly constructing the rest of the theme with the teacher.

4.2.2 Integrating content and language by combining thematic patterns and genre structures

Apart from providing designed and spontaneous scaffoldings to help students understand the thematic patterns and semantic relationships, the teacher also reminded students of the language knowledge which is inseparable from meaning making of the content knowledge. For example, during answer-checking, a student mixed up "*stable*" and "*strong*" when describing the characteristics of the DNA molecule. Miss T helped him select the appropriate modifier by providing spontaneous scaffolding about the thematic pattern.

Excerpt 2. Content and language integration based on thematic patterns

T	*How about the third sentence? How about the third sentence? LG. How about the third sentence? Third.*
LG	*DNA is a strong molecule because…*
T	*Is a … What molecule?*
LG	*Strong.*
T	*Is a strong molecule? You use strong here? Um? YY, would you help him?*
YY	*DNA is a* stable *molecule.*
	Third, DNA is a stable molecule
	[T showing the third characteristic on the screen.]
T	*Is a stable molecule. Stable is better than strong here. Okay. You can say that the bonding is very strong. The covalent bond is very strong but the hydrogen bond is relatively weaker. Okay. But we won't say that the molecule is strong. The molecule is stable. YY. Sit down please.* [Turning to LG again and ask another question.] *Why the DNA molecule is stable?*
LG	*Because it has strong* sugar-phosphate *backbones.*

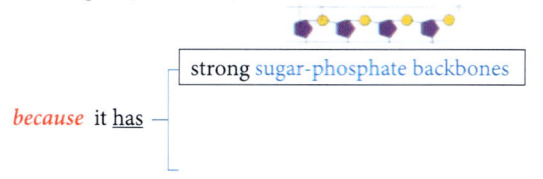

            strong sugar-phosphate backbones
     because it has

[T showing the words about the reason on the screen.]

T	*Good! It has strong sugar-phosphate backbones.*

In this example, the collocations of the two adjectives (EPITHET: stable and strong) with the corresponding nouns (THING: molecule and backbone) were related to the specific characteristic of the DNA structure and its function which implied a cause and consequence relationship; i.e., DNA is a *stable* molecule because it has *strong* sugar-phosphate backbones and a double-helix structure. The students needed to understand the semantic relationship in the thematic pattern to decide the proper modifier. Hence, it was necessary for the teacher to supplement this spontaneous scaffolding about the relationship between thematic patterns and language features to help students clarify the proper collocations between modifiers and things in the thematic patterns which enabled them to understand the characteristic of DNA accurately.

During teacher-student interactions, spontaneous scaffolding on language knowledge were also provided intermittently. For example, when discussing the first theme, Miss T asked students to use the synonyms "consists of", "contain" and "has" when they expressed the meaning of composition which appeared repeatedly in the C+L map. Similar examples also included the paraphrasing of "passed on" and "transmit" which she encouraged students to associate the learned vocabulary (e.g., transmission) with the newly appear ones (e.g., pass on). When the teacher elaborated on the key concept "complementary base pairing", she asked a student to use an example from the content knowledge to explain "complementary" (*"What does complementary mean? Would you give us an example? If the base is A, it should pair with…?"*). By doing so, Miss T conveyed to students a message that language use was closely associated with the meaning network – thematic patterns of the science lesson. She also reminded students to use academic vocabulary (e.g., replicate) to replace the everyday words (e.g., copy) when answering essay questions which was a typical weakness of the students in high-stake exams.

5. Discussion

The research findings indicated that the thematic-pattern-based CLM approach facilitated both content and language knowledge development in the EMI biology class. Data analysis revealed that concept and language mapping was a process of integrating content and language by intertwining thematic-pattern-based (Lemke, 1990) designed and spontaneous scaffoldings (Gibbons, 2009; Lin, 2016) in dialogic/authoritative interactions (Mortimer & Scott, 2003). The CLM pedagogy started with the design of CLM materials and activities according to the thematic topic of the content subject. The thematic patterns and semantic relationships were represented by the multimodal animated CLM materials in forms of C+L cards, C+L maps, sentence-making tables and essay writing guides which were circulated in the EMI biology lessons through a series of talking, reading, writing, representing and doing communication activities (Osborne, 2014) in the "repetition with variation" strategy (Lemke, 1990).

The CLM materials as designed scaffolding make science learning more *"focused"* and *"complete"* by condensing and weaving thematic relations into thematic nexus. What make the CLM pedagogy more *"flexible"* and *"impressive"* are the multimodal (e.g., the diagrams besides the corresponding thematic items in C+L cards/maps), animated (e.g., the dynamic emerging of thematic items one by one to keep teaching and learning at the same pace) and sequential (e.g., the *"from simple to complex"* arrangement of thematic patterns) design of the C+L maps on the PowerPoint slides. All these "properties" are best orchestrated through the spontaneous scaffolding of the teacher who guides the joint construction of the science story through a series of triadic dialogues based on not only the thematic patterns of the science lesson but also the academic language features in the science texts. Following the "repetition with variation" principle, the thematic patterns about the same thematic topic which are explicitly introduced in the same lesson or implicitly related to previous lessons will be connected, talked about, and further explored through a series of C+L activities. It should be noted that, in EMI CLIL context where the content subject is taught in an additional language of the students (teachers), the combination of thematic patterns and rhetorical/genre structures is equally crucial for effective science teaching.

5.1 Integrating content and language in CLIL lessons

Previous studies exploring CLIL have accentuated the integration between content and language as the "core concept" (Lorenzo, 2016). Morton and Llinares (2017) emphasized the need to clarify the "actual meaning of the label" (i.e. integration) pointing out the phenomenon that "…the term (CLIL) seems to be mainly used to describe bilingual education context where content classes are taught through an additional language but where little integration of content and language happens" (p. 2). Lyster (2007) proposed the counterbalanced approach, but called for further research on the link between the subject-matter class and the English-as-foreign-language class. Lin (2016) also stressed that a well-developed framework for description of language patterns is available (e.g., the Genre-Egg framework based on the Sydney School of genre theories), "however, for the precise description of content, we still need to develop a theoretical framework to enable us to describe units of meaning in specific content areas" (p. 179).

The thematic-pattern-based CLM pedagogy proposed in this study to some extent achieved the effect of content and language integration. As Lemke (1990) clarified, "Talking Science' does not mean simply talking about science; it means doing science through the medium of language" (Abstract of *Talking Science*). According to Lemke, students should be taught both the thematic patterns and the genre of science because reasoning is based on both the use of thematic patterns and genre structure patterns, the former supplies the content and the latter supplies the form of organization of the argument. As can be seen from the data analysis, the CLIL lessons in the CLM approach cannot separate content from language, as the thematic patterns (i.e., "a network of relationships among the scientific concepts in a field, but described semantically, in terms of how language is used in that field" (Lemke, 1990, p. 12)) in forms of CLM materials and activities linked up every teaching stage of the CLIL lessons with thematic-pattern-based designed scaffolding and spontaneous scaffolding facilitating the talking, reading, representing, writing and doing science in the CLIL classroom.

5.2 Drawing on perspectives of subject education researchers

Echoing Halliday's (1993) point that content and language are always integrated, researchers and educators in subject education also base their work on the premise that "learning the language of science is a major part (if not the major part) of science education. Every science lesson is a language lesson" (Wellington & Osborne, 2001, p. 2). In this study, not only do science experts provide us with important theoretical implications (e.g., thematic patterns theory by Lemke, 1990; concept mapping by Novak et al., 1983; and the review of scientific practices and inquiry by Osborne, 2014), they also propose useful pedagogical strategies and techniques from their research findings. For example, the design of multimodal animated C+L materials has adopted the research designs of subject education researchers such as Cheng and Gilbert (2015) and Nesbit and Adesope (2011). The feedback of the teacher and students as well as the lesson observation findings also proved that the pedagogical techniques proposed by the subject experts were important scaffoldings for CLIL lessons.

5.3 Teacher education about "thematic-pattern-based" CLM pedagogy

The thematic-pattern-based CLM approach being a newly developed pedagogy, its feasibility, practicality and sustainability need to be carefully discussed and reflected on. Judging by the background of the EMI curriculum in secondary education in Hong Kong and many other regions and countries where content subjects are taught in an additional language of the students (and the teachers), the CLM approach may contribute to CLIL practices by providing a potentially feasible pedagogy and research-based references. However, it should be noted that, under exam-oriented school culture, there may be challenges in adapting this new pedagogy to EMI classrooms. Some students have become so accustomed to rote learning that they may find the CLM materials contain "not enough words" for them to memorize directly; for example, one student explained that she recited all notes word by word because she would learn science subjects as if she was learning Chinese History, a subject which students believed involves considerable amount of memorization. Although the majority of students had positive feedback on the CLM pedagogy, some of them worried about the limit of lesson time, as one commented, on the one hand the teacher's step by step guidance based on the C+L maps as well as her questioning following the animated sequencing of the different bits of information helped him to learn the concepts better with clearer understanding; on the other hand he also thought the pedagogy time consuming as the teacher could just give students the answers and ask them to check the worksheet themselves. To address these challenges, teachers adopting the CLM approach need to emphasize two principles: first, rather than learning by rote memorization, the pedagogy encourages learning by meaning making; namely, to help students to integrate content and language learning through understanding the semantic relationships within and between the thematic patterns of content subjects; second, the pedagogy does not just provide CLM materials as designed scaffolding, more importantly, it accentuates the spontaneous scaffolding (Gibbons, 2009; Lin, 2016) – the teacher's step by step deployment of the "concept + language mapping" materials and activities as well as the classroom interactions during which the teacher guides the students to achieve thematic coherence (Bloome, Carter, Christian, Otto, & Shuart-Faris, 2005) of the CLIL lessons.

Another key issue regarding teacher education will be the teacher knowledge about the CLM approach. Morton (2018) re-conceptualized CLIL teacher knowledge and proposed the construct of "language knowledge for content teaching" with two sub-domains: common language knowledge for content teaching and specialized language knowledge for content teaching. The thematic-pattern-based CLM approach demands not only teacher language knowledge and content knowledge, but how the two types of knowledge may be integrated according to the CLIL lesson. It would be unrealistic for teachers to self-learn the theory and develop the CLM materials all at once as they may feel too abstract to fully understand the theory only by self-directed learning. The "Collaborative, Dynamic and Dialogic Process" CLIL teacher education model (He & Lin, 2018) may be one of the solutions. Teachers may join Master of Education programmes in the field of CLIL or sharing sessions of CLIL teacher professional development workshops to learn about the theory and skills relevant to the CLM pedagogy and then try out part of the CLIL lessons by collaborating with colleagues in the school.

6. Conclusion

In this study, we developed a thematic-pattern-based CLM approach and tried it out in an EMI biology classroom. Both quantitative and qualitative data indicated that the CLM approach had a positive effect on students' development of both content and language knowledge. However, it should be noted that the present study adopted a quasi-experimental design but there was only one intervention class and one control class with around 30 students in each cohort. The limit of class number and class size may affect the quantitative result of the study. Future research on thematic-pattern-based CLM approach may need to increase the number of classes and adopt a longitudinal research design. Intervention may be tried out in other subjects with medium of instruction other than English. Data collection may also include students' design and elaboration on their own CLM materials, e.g., how students express their understanding of the thematic patterns through their own C+L maps. Data analysis may focus on the effects of interactive/dialogic communications on students' content and language development.

Judging by the shortage of evidence-based research on CLIL and the difficulties that CLIL teachers have encountered (e.g., lack of pedagogical support and CLIL teacher education, tight teaching schedule, heavy workload, pressure of high-stake exams, etc.), we recommend more support for the research of CLIL education and CLIL teacher professional development (He & Lin, 2018). The thematic-pattern-based CLM pedagogy, research methods as well as research findings of this study may be useful resources upon which further investigation can be developed.

Acknowledgements

The paper is based on data from a large project funded by the Standing Committee on Language Education and Research (SCOLAR) (Project #2015–0025) awarded to Angel M.Y. Lin. We are grateful to the participating teachers and students for their support. We also thank the anonymous reviewers for their suggestions.

References

Ausubel, D. P. (1968). *Educational psychology: A cognitive view*. New York, NY: Holt, Rinehart and Winston.
Bloome, D., Carter, S. P., Christian, B. M., Otto, S., & Shuart-Faris, N. (2005). *Discourse analysis and the study of classroom language and literacy events. A micro ethnographic perspective*. Mahwah, NJ: Lawrence Erlbaum Associates.

Cammarata, L., & Ó Ceallaigh, T. J. (2018). Teacher education and professional development for immersion and content-based instruction: Research on programs, practices, and teacher educators. *Journal of Immersion and Content-Based Language Education*, 6(2), 153–161. https://doi.org/10.1075/jicb.00004.cam

Cammarata, L., & Tedick, D. (2012). Balancing content and language in instruction: The experience of immersion teachers. *Modern Language Journal*, 96(2):153–289. https://doi.org/10.1111/j.1540-4781.2012.01330.x

Cenoz, J. (2016). Discussion: Towards an education perspective in CLIL language policy and pedagogical practice. In Y. Ruiz de Zarobe (Ed.), *Content and language integrated learning: Language policy and pedagogical practice*. London: Routledge.

Cenoz, J., Genesee, F., & Gorter, D. (2014). Critical analysis of CLIL: Taking stock and looking forward. *Applied Linguistics*, 35(3), 243–262. https://doi.org/10.1093/applin/amt011

Cheng, M. M. W., & Gilbert, J. K. (2015). Students' visualization of diagrams representing the human circulatory system: The use of spatial isomorphism and representational conventions. *International Journal of Science Education*, 37(1), 136–161. https://doi.org/10.1080/09500693.2014.969359

Coyle, D., Hood, P., & Marsh, D. (2010). *CLIL: Content and language integrated learning*. Cambridge: Cambridge University Press.

Creswell, J. W. (2003). *Research design: Qualitative, quantitative, and mixed methods approaches* (2nd ed.). Thousand Oaks, CA: Sage.

Dale, L., Ron, O., & Verspoor, M., (2018). Searching for identity and focus: Towards an analytical framework for language teachers in bilingual education. *International Journal of Bilingual Education and Bilingualism*, 21(3), 366–383. https://doi.org/10.1080/13670050.2017.1383351

Dalton-Puffer, C. (2013). A construct of cognitive discourse functions for conceptualizing content-language integration in CLIL and multilingual education. *European Journal of Applied Linguistics*, 1(2), 216–253. https://doi.org/10.1515/eujal-2013-0011

Dalton-Puffer, C. (2018). Postscriptum: Research pathways in CLIL/Immersion instructional practices and teacher development. *International Journal of Bilingual Education and Bilingualism*, 21(3), 384–387. https://doi.org/10.1080/13670050.2017.1384448

Dalton-Puffer, C., & Nikula, T. (2014). Content and language integrated learning. *The Language Learning Journal*, 42(2), 117–122. https://doi.org/10.1080/09571736.2014.891370

Gibbons, P. (2009). *English learners, academic literacy, and thinking: Learning in the challenge zone*. Portsmouth, NH: Heinemann.

Halliday, M. A. K. (1993). Towards a language-based theory of learning. *Linguistics and education*, 5(2), 93–116. https://doi.org/10.1016/0898-5898(93)90026-7

He, P. C., & Lin, A. M. Y. (2018). Becoming a "language-aware" content teacher: Content and language integrated learning (CLIL) teacher professional development as a collaborative, dynamic, dialogic process. *Journal of Immersion and Content-Based Language Education*, 6(2), 163–189. https://doi.org/10.1075/jicb.17009.he

Lemke, J. L. (1990). *Talking science: Language, learning and values*. Westport, CT: Ablex.

Lemke, J. L. (1998). *Teaching all the languages of science: Words, symbols, images, and actions*. Conference on Science Education in Barcelona.

Lin, A. M. Y. (2007). What's the use of 'triadic dialogue'? Activity theory, conversation analysis and analysis of pedagogical practices. *Pedagogies*, 2(2), 77–94. https://doi.org/10.1080/15544800701343943

Lin, A. M. Y. (2016). *Language across the curriculum & CLIL in English as an Additional Language (EAL) Contexts: Theory and practice*. Dordrecht: Springer. https://doi.org/10.1007/978-981-10-1802-2

Lin, A. M. Y., & Man, E. Y. F. (2009). *Bilingual education: Southeast Asian perspectives*. Hong Kong: Hong Kong University Press. https://doi.org/10.5790/hongkong/9789622099586.001.0001

Llinares, A., Morton, T., & Whittaker, R. (2012). *The roles of language in CLIL*. New York, NY: Cambridge University Press.

Lorenzo, F. (2016). Genre-based curricula: multilingual academic literacy in content and language integrated learning. In Y. Ruiz de Zarobe (Ed.), *Content and language integrated learning: Language policy and pedagogical practice*. London: Routledge.

Lyster, R. (2007). *Learning and teaching languages through content: A counterbalanced approach*. Amsterdam: John Benjamins. https://doi.org/10.1075/lllt.18

Mortimer, E., & Scott, P. (2003). *Meaning making in secondary science classrooms*. Maidenhead: Open University Press.

Morton, T. (2018). Reconceptualizing and describing teachers' knowledge of language for content and language integrated learning (CLIL). *International Journal of Bilingual Education and Bilingualism*, 21(3), 275–286. https://doi.org/10.1080/13670050.2017.1383352

Morton, T., & Llinares, A. (2017). Content and Language Learning (CLIL): Type of programme or pedagogical model? In A. Llinares & T. Morton (Eds.), *Applied linguistics perspectives on CLIL* (pp. 105–24). Amsterdam: John Benjamins. https://doi.org/10.1075/lllt.47.01mor

Nesbit, J. C., & Adesope, O. O. (2011). Learning from animated concept maps with concurrent audio narration. *The Journal of Experimental Education*, 79, 209–230. https://doi.org/10.1080/00220970903292918

Nikula, T., Dafouz, E., Moore, P., & Smit, U. (2016). *Conceptualizing integration in CLIL and multilingual education*. Bristol: Multilingual Matters. https://doi.org/10.21832/9781783096145

Novak, J. D. (2010). *Learning, creating, and using knowledge: Concept maps as facilitative tools in schools and corporations* (2nd ed.). New York, NY: Routledge. https://doi.org/10.4324/9780203862001

Novak, J. D., Gowin, D. B., & Johansen, G. T. (1983). The use of concept mapping and knowledge vee mapping with junior high school science students. *Science Education*, 67(5), 625–645. https://doi.org/10.1002/sce.3730670511

Osborne, J. (2014). Scientific practices and inquiry in the science classroom. In Lederman, N. G. & Abell, S. K. (Ed.), *Handbook of research on science education* (Volume II). New York, NY: Routledge.

Reeves, T. C. (2000). Enhancing the worth of instructional technology research through "design experiments" and other development research strategies. *International perspectives on instructional technology research for the 21st century*, New Orleans, LA.

Rose, D., & Martin, J. R. (2012). *Learning to write, reading to learn: Genre, knowledge and pedagogy in the Sydney School*. Sheffield: Equinox.

Ruiz de Zarobe, Y. (2016). Introduction – CLIL implementation: From policy-makers to individual initiatives. In Y. Ruiz de Zarobe (Ed.), *Content and language integrated learning: Language policy and pedagogical practice*. London: Routledge.

Wellington, J. J., & Osborne, J. (2001). *Language and literacy in science education*. Buckingham: Open University Press.

Appendix 1. Summary of student feedback on the CLM materials

C+L cards

S1 *I use the C+L cards to help me memorize the concepts… Because Miss T used to end a unit very quickly, to understand the lessons better, we need to highlight the key points by ourselves and then read them several times after class. In fact, many key points are already there (in the C+L cards) and we simply glance at the cards and get the idea of the concepts…I think they (C+L cards) are quite helpful. Actually, sometimes, even though I'll highlight the key points myself, I would miss some or just skip them directly.*

S2 *I think it'll be much better if we have the C+L cards, because, frankly speaking, textbooks usually have many useless words, I mean, words that are not so relevant. The C+L cards help us summarize all key points without any nonsense.*

S5 *I find the notes in the C+L cards helpful for me, because I also make notes myself, and I used to make notes in point forms. Comparing the C+L card notes with mine, now I'll make them more detailed; like, I'll add some pictures and diagrams to make them more impressive.*

S2 *I think the C+L cards are better (than the bullet point notes in the textbook) because their ideas are in complete sentences showing an entire point in a whole process; but the textbook bullet points only mention the key parts without telling what happened before and what will come next.*

S3 *Yeah, I like them (diagrams, arrows and boxes on the C+L cards), because sometimes if I just read the text, it'll be too hard for me to imagine the concepts at once, but if the diagrams are shown, I can visualize the concepts right away. For example, if the diagram of a cell's epithelium is shown, I can think of its functions at once.*

Ss *When doing exercises, will you take out the corresponding C+L cards to see, for example, what are "recessive" and "dominant", and how to distinguish "heterozygous" and "homozygous"? Will you?*
Yes.

	C+L maps
S5	I noticed that some words are deliberately bold and some underlined (in the C+L maps) ... I'll pay particular attention to these words, and I think it really makes learning easier because the concepts can be put one by one back into the C+L Map.
S1	Miss T raises questions (about the concepts in the C+L map), students look up answers from the C+L cards, textbook and handouts, and then the teacher goes on elaborating on concepts in the next layer. Is it good to follow such a way to help you consolidate your knowledge? There are pros and cons. Oh, why? The cons is that it slows down the teaching, we should have remembered what the answer is as the teacher has just explained it. But for the pros, it makes learning more impressive, because the teacher raises a question (about a concept) and we think about it and then answer the question (with the help of the C+L materials). Such Qs and As help us understand the concepts better.
Ss	If we collaborate with Miss T again, is it good to design more such C+L maps?
S2	Yes. Because they're simple and easy to understand.

	Sentence-making tables
S4	I find them (sentence-making tables) quite useful, because we've learned more verbs, especially some subject-specific ones which are must-use words in the unit. So I think it makes things easy.
S3	I think they (sentence-making tables) are quite good, because, for example, the cause and effect relation, everything has its cause and effect, the tables illustrate this very clearly. They also have some subject-specific terms, like "replicate", it can't be replaced by "copy". The tables highlight these clearly so that we can learn them well.
Ss	If you are doing a test, such as an essay question, do you know how to use the sentence patterns in the sentence-making tables to answer the questions? (Nodding showing understanding)
S5	I find the sentence-making tables helpful, because, like this one, you can see the definition at once, just like what you said, there is a pattern telling you (how to use it).

Scaffolding for cognitive and linguistic challenges in CLIL science assessments

Yuen Yi Lo,[1] Wai-mei Lui[2] and Mona Wong[1]
[1] The University of Hong Kong [2] The University of Alberta

In Content and Language Integrated Learning (CLIL) programmes, students learn some non-language content subjects through a second/foreign language (L2), and their content knowledge is often assessed in their L2. It follows that students are likely to face challenges in both cognitive and linguistic aspects in assessments. Yet, there has been limited research exploring whether and how CLIL teachers help their students cope with those challenges. This multi-case study seeks to address this issue by investigating the instructional and assessment practices of two science teachers in Hong Kong secondary schools. The two teachers presented an interesting contrast – one teacher incorporated both implicit and explicit language instruction in her lessons, so her students were well prepared for the assessment tasks; the other teacher's instructional and assessment practices were heavily content-oriented, and it is not sure whether students mastered both content and L2. These findings illuminate CLIL pedagogy and teacher education.

Keywords: bilingual education, Content and Language Integrated Learning (CLIL), instruction, assessment

1. Introduction

Content and Language Integrated Learning (CLIL) is defined as "any type of pedagogical approach that integrates the teaching and learning of content and second/foreign languages" (Morton & Llinares, 2017, p. 1). Although the term CLIL was coined in Europe in the 1990s to refer to increasingly popular attempts to use students' second/foreign language (L2), very often English, as the medium of instruction in non-language content subjects, it is now regarded as an umbrella term to encompass different bilingual programmes following a similar principle of content and language integration (e.g., immersion programmes in Canada, English as the medium of instruction [EMI] education in Asia) (Cenoz, Genesee, &

Gorter, 2014). CLIL is also adopted as an umbrella term in this paper, and our research context of EMI secondary schools in Hong Kong is considered as a variant of CLIL, although EMI is often placed towards the content-driven end of the continuum illustrating different bilingual programmes (Lin, 2016).

The worldwide spread of CLIL to different educational contexts, especially to those English as a foreign language (EFL) ones, has attracted a great deal of research efforts, which to date have largely focused on student achievements (e.g., Admiraal, Westhoff, & de Bot, 2006; Navés, 2011) and classroom interaction and discourse (e.g., Lin & Wu, 2015; Morton & Jakonen, 2016). The important issue of assessment in CLIL is under-explored (Hönig, 2010; Massler, Stotz, & Queisser, 2014). In CLIL, students are usually assessed of their content knowledge in the L2, which is a language they are still acquiring and are less proficient in. Hence, students are actually facing challenges in both cognitive and language aspects when they attempt CLIL assessments (Lo & Fung, 2020). In this sense, it would be essential to examine whether and how teachers provide scaffolding for both aspects, so that the assessment data collected can reflect students' learning progress. Moreover, it has been argued that assessment has "backwash effect" on teaching and learning behaviours (Alderson & Wall, 1993). Hence, delving into the assessment issues in CLIL can inform what teachers and students have to do in order to move towards their learning targets in terms of content and language.

This study seeks to contribute to the under-researched area of CLIL assessments by examining the alignment among objectives, instruction and assessments, so as to better understand whether and how CLIL teachers prepare their students to tackle the challenges in assessments. Its findings will provide significant implications for effective pedagogical practices and teacher education in CLIL. The complexity of aligning objectives, instruction and assessments in CLIL will first be discussed, which will pinpoint the importance of providing scaffolding for students.

2. Literature review

2.1 Alignment among objectives, instruction and assessments in CLIL

In assessment and testing literature, "validity" is an essential element when developing educational measurement, as it concerns whether a particular test is measuring what it is intended to measure (Hughes, 2003). It is argued that a valid assessment should be aligned with programme/lesson objectives and classroom instruction (Orlich, Harder, Callahan, Trevisan, & Brown, 2013). If students are not assessed on what they have been taught, then the assessment cannot provide useful information about students' progress and teaching effectiveness. When such alignment is applied to CLIL, it becomes more complicated. As depicted in Figure 1, the programme/lesson objectives should include both content and language objectives, which are the dual goal of the programme; the instructional activities should be planned to achieve the objectives in both dimensions, by counterbalancing the lesson orientation between content and language (Lyster, 2007) or by incorporating language teaching into the usually more content-oriented CLIL lessons (Llinares Morton, & Whittaker, 2012); and the assessment tasks should also consider both content/cognitive and language dimensions.

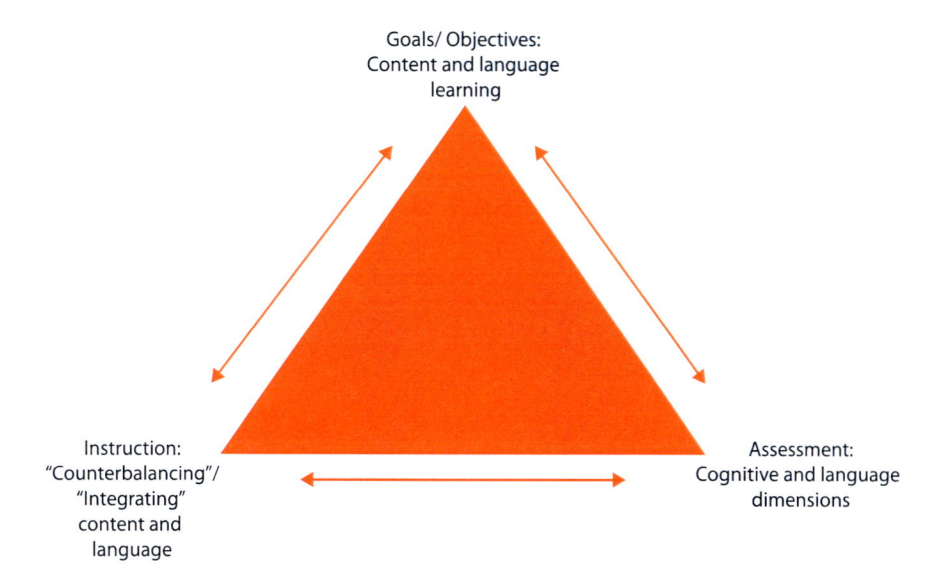

Figure 1. Alignment among programme/lesson objectives, instruction and assessment in CLIL

2.2 CLIL objectives and teachers' instruction

Given the dual focus on content and language teaching in CLIL, all teachers are responsible for teaching language to a certain extent (Cammarata & Haley, 2018; Llinares et al., 2012). The CLIL programmes implemented in different educational contexts may have different practices and teacher and student profiles. In those contexts where CLIL is mainly practised in content subject lessons (e.g., in EMI schools in Hong Kong), the CLIL teachers have usually been trained as subject specialists and they tend to pay more attention to content teaching (Lo, 2014; Tan, 2011). Cammarata's (2016) framework for planning content-language-literacy integrated curricula proposes that CLIL teachers need to pay attention to "content objectives", "content-related language objectives", "academic literacy skills objectives" and "literacy-related language objectives". However, it has been reported that CLIL teachers encounter great difficulties in identifying language objectives during lesson planning, and they focus primarily on vocabulary teaching in their lessons, which seems to suggest that they are not highly aware of other features of academic language (e.g., grammar, sentence structures, genres) (Baecher, Farnsworth, & Ediger, 2014; Cammarata & Haley, 2018). In terms of instruction, quite a few studies have observed that CLIL subject teachers know little about CLIL teaching strategies (Kong, Hoare, & Chi, 2011), second language acquisition theories (Koopman, Skeet, & de Graaff, 2014), or principles for material adaption and design (Pérez-Cañado, 2016). In view of this, CLIL researchers have proposed some frameworks to help CLIL teachers to better incorporate language teaching into their content-oriented lessons. In addition to Cammarata's (2016) framework for curriculum planning aforementioned, Lyster (2016) put forward an integrated instructional sequence, with four phases namely noticing of the target language features in the context provided by content subjects, awareness and metalinguistic reflection of the target features, guided practice and autonomous practice of using the target features in meaningful content-related contexts. Llinares et al. (2012) and Lo and Jeong (2018) have explored the implementation of genre-based pedagogy as a possible pedagogical framework to help students develop academic literacies.

2.3 CLIL objectives and assessment

When designing assessments in CLIL, one core question to ask is "what to assess?" – whether the focus should be on content or language, or on both (Coyle, Hood, & Marsh, 2010). Theoretically speaking, both content and language should be assessed, so that they align with the dual goal in CLIL (Massler et al., 2014). Yet, it is very difficult to do so, considering the fact that some CLIL is often practised in content subjects and hence content curriculum tends to determine the parameters for assessments (Hönig, 2010). Reierstam (2015) analysed the perceptions and assessment practices of biology and history teachers in CLIL and non-CLIL programmes in Sweden, and she observed that CLIL teachers tended to put more emphasis on assessing "content", whereas "language" meant only subject-specific terminology to them. In a similar vein, the CLIL history teachers in Austria in Hönig's (2010) study placed greater emphasis on students' content knowledge when they designed assessment tasks and marking rubrics. However, they were actually examining both content and language implicitly because students' language proficiency (e.g., accuracy) did influence their marking. Thus, it is worth further exploring how CLIL teachers design assessment tasks that consider both content and language dimensions.

2.4 CLIL assessment and teachers' instruction

Assessment clearly has an impact on classroom practices, which is often referred to as the "backwash effect" (Alderson & Wall, 1993). As aforementioned, content curriculum goals dominate the assessment of content subjects, so CLIL subject teachers may not be aware of the need to incorporate more language teaching into their lessons (Tan, 2011). At the same time, the role of language in CLIL assessment is highly complex, and several issues need to be considered. First, it has been shown that students can better express their content knowledge in their first language (L1) than in their L2 in research requiring them to complete the same task in both languages (Gablasova, 2014). Hence, assessments in CLIL bear the risk of underestimating students' actual knowledge in content subjects. After examining the question papers and student scripts of an internationally recognised examination (IGCSE) taken by numerous non-native English students, Shaw and Imam (2013) observed that students with insufficient linguistic resources failed to get maximum marks on questions requiring more developed answers (e.g., essays in geography and history papers). In another study, Lo and Fung (2020) surveyed the questions in science/biology textbooks and public examination papers in Hong Kong EMI education. They analysed all the questions with regard to the cognitive and linguistic demands imposed on students. Their findings demonstrated the integral role played by language in CLIL assessments, as most of the times, students were required to understand the questions presented in sentences or short texts, and they were also expected to express their content knowledge through sentences or short texts. Such findings raise the concern about whether CLIL teachers are equipping their students with the knowledge and skills to overcome the challenges in cognitive and linguistic dimensions in assessments. Otherwise, the data collected from CLIL assessments may not be valid, in the sense that they are not reflecting what the CLIL teachers have taught (which tends to be heavily content-oriented), nor what the students have learned (as they may have been hindered by language barriers).

In recognition of the above important yet unanswered issues in CLIL assessments, this study seeks to investigate the relationship among programme/lesson objectives (i.e., content and language integrated learning), classroom instruction and assessment practices. In particular, the study aims to examine whether and how CLIL teachers provide scaffolding to prepare students to tackle both content and language challenges in assessments, thereby achieving the dual goal of the programme. "Scaffolding" is an important concept in the social constructivist view of learning (Gibbons, 2015; Woods, Bruner, & Ross, 1976). It refers to the process that enables a learner to accomplish a task which would otherwise be beyond his/her efforts. This paper focuses on instructional scaffolding, which includes any kind of support that teachers provide during instruction, such as oral explanation, instructional tasks or activities and demonstration. And these scaffolds include linguistic support (which aims at making the target language more accessible) and conceptual support (which assists students in better understanding the concepts) (Pawan, 2008).

With such aims in mind, this study is asking *"To what extent does teachers' instruction in lessons align with the goals of content and language integrated learning, as well as assessment practices?"*

3. Methodology

3.1 Overall research design

This study adopted a multi-case study approach, which allows holistic and in-depth investigation of a phenomenon in its real-life context (Yin, 2009). One CLIL teacher constituted one case, and multiple cases were included to investigate a variety of school contexts and teachers and to enhance the external validity of the findings. In what follows, the research context and participants will first be described, and then more details about data collection and analysis will be presented.

3.2 Research context and participants

The study was conducted in Hong Kong, where EMI education has been in place for decades. Owing to its colonial history and socio-economic development, while Chinese (spoken Cantonese and Standard Written Chinese) is the language for daily communication among the majority of population, English is regarded as an important social capital for one's academic and career prospects (Li, 2017). Hence, there has always been a strong preference (from parents) for EMI education, which normally takes place at secondary level (Grade 7 to 12; 12 to 18 years old). Under the government's most recent medium of instruction policies, most secondary schools in Hong Kong can teach at least some content subjects in English. In other words, CLIL, used as an umbrella term in this paper, is being implemented in most secondary schools in Hong Kong, though to different extents. Such diversified practices of CLIL provide a favourable context for this study, which took into consideration different factors (including the school context) when examining the alignment among objectives, instructional practices and assessments. As aforementioned, the EMI education in Hong Kong is content-driven, and the explicit learning goals are content-oriented (i.e., focusing on students' content knowledge and cognitive skills). A quick survey of the curriculum documents in Hong Kong conducted by the research team did not identify any explicit emphasis on "(English) language teaching or learning". As the same curriculum documents apply to both Chinese-medium and EMI education, the researchers could only find such terms as "reading to learn", "communication" and "communicative skills", which are related to language teaching and learning.

Twelve teachers, including five teaching science, were recruited for the study. These teachers came from schools in different districts and had different years of teaching experience and professional training. As will be illustrated below, various sources of data were collected from each of these teacher participants. The whole data collection process for the project lasted from September 2016 to June 2017, with the research team spending around two to three weeks with each case teacher on average. The space in one paper does not allow us to discuss all the cases, especially when we seek to present a more comprehensive picture of the teachers' instruction and assessment practices. Therefore, we will focus on reporting two science teachers, who are considered as illustrative cases in this study, since they represent the two major patterns of practices observed in this study. The details of the two selected science teachers will be presented below.

Both Miss A and Miss B taught science and were experienced, with over 15 years of teaching experience. Their teaching qualifications were similar – they were trained as science teachers and had received some in-service training related to CLIL. Perhaps the major difference lay in their school contexts – Miss A was teaching in a top school where all subjects (except Chinese-related ones) were taught in English, whereas Miss B was teaching in an average school where at junior secondary levels, some classes only learned science and mathematics through English. Hence, it would be reasonable to assume that in general, the academic ability and English proficiency of students in Miss A's class were higher than those in Miss B's class. The profiles of the two cases are summarised in Table 1.

3.3 Data collection

3.3.1 *Lesson observations*

To understand the teachers' instructional practices, each teacher was observed when teaching one unit of the subject. These lessons were video- or audio-recorded, and at least one research team member was present to jot down field

Table 1. Profiles of the two cases

	Miss A	Miss B
Years of teaching experience	>20	16
Teaching qualifications	Subject trained; with teacher qualification; some in-service training on CLIL/EMI	
School context	– Band 1 top girls' school – All subjects (except Chinese-related ones) were taught in English	– Band 2 average co-educational school – Only science & mathematics were taught in English
Grade level observed	Grade 9	Grade 8
Number of lessons & Topic involved	4 lessons on "Application of Enzymes" (practical included)	4 lessons on "Common Acids and Alkalis" (practical included)

Note.
* Hong Kong primary school leavers are categorised into three bands, with band 1 being the highest band. The categorisation is conducted based on students' performance on the subjects of Chinese, English, Mathematics and General Studies, in both internal and external examinations.

notes, according to a lesson observation protocol prompting notes about lesson objectives, instruction related to content and/or language, as well as assessment practices (attached in Appendix 1). Four lessons taught by Miss A and Miss B were observed and recorded respectively. These yielded a total of 357 minutes (nearly 6 hours) of lesson recording for analysis. The researchers also managed to have a brief chat with the teachers before each of the lessons observed, so as to understand their lesson plan and objectives.

3.3.2 *Collection of assessment tasks*

The formative and summative assessment tasks, together with a random sample of marked scripts, for the unit observed were collected by each teacher. Yet, the time lag between lesson observations and summative assessment period often resulted in a discrepancy in the assessment tasks collected. For example, the researchers could only gather a formative written assessment task from Miss A and 10 scripts from her students. From Miss B's class, the researchers managed to collect both formative and summative assessment tasks, and 12 sample scripts. To make the illustration of the two cases more comparable, this paper will focus only on the two teachers' formative assessment practices. In this paper, formative assessment tasks included in-class worksheets/practices and homework, all of which offer opportunities for teachers to understand students' progress and provide feedback to promote further learning (see, for example, Earl [2013], for a more thorough discussion about formative and summative assessment). It is acknowledged that questions asked by teachers during instruction could also serve as formative assessment by checking students' understanding. However, owing to the limited space in this paper, this type of formative assessment is not included.

3.3.3 Semi-structured interviews

The perceptions of both teachers and students of the classroom practices and assessment in EMI education were collected, so as to triangulate or elaborate on what was observed in lessons and analysis of assessment tasks. An individual semi-structured interview was conducted with each case teacher, whereas focus group interviews were conducted with one group of students (3–4 students in a group) from each observed class. The interviews with the teachers lasted for around 45 minutes each, whereas those with the students lasted for around 30 minutes. The student interviewees were nominated by the teachers, who were asked to select students with different levels of academic ability. All the interviews were conducted in Cantonese, the participants' first language.

3.4 Data analysis

All the observed lessons and interviews with teachers and students were transcribed verbatim to allow for detailed analysis. These, together with the documents collected, were analysed and coded for recurrent themes related to the research question (i.e., the alignment among objectives, instruction and assessment practices; teachers' scaffolding).

In particular, the objectives of the unit and lessons being observed were inferred from the pre-lesson observation chats with the teachers and sometimes from what the teachers told the students during the lessons. These would reveal the attention paid by the teachers to content and language teaching.

The transcribed lessons were analysed according to their foci, based on a coding scheme devised by the research team (see Figure 2). First, a lesson was divided into various "episodes" which focused on a particular theme or served a particular function (e.g., greetings and introducing the lesson; review of the previous lesson; instruction of a concept). These episodes were then classified into instructional register (those focusing on knowledge delivery and discussion) or regulative register (those managing classroom tasks and students' behaviours) (Christie, 2002; see Appendix 2 for examples of different registers). Second, those episodes under instructional register were further classified according to their focus on "content" or "language", and then their respective level. The three levels for "content" include "recall", "application" and "analysis", which were condensed from the

Bloom's taxonomy (Krathwohl, 2002), to represent the different levels of cognitive challenges. The three levels for "language" are "lexico-grammar", "sentence" and "text", which largely correspond to the different levels of language features when analysing academic language (Lin, 2016). The analysis generated by such coding procedures can reveal the attention paid to content and language in the lessons observed (i.e., whether a particular lesson tended to be more "content-oriented" or "language-oriented").

Major theme	Sub theme	
Instructional register	Content	Recall
		Application
		Analysis
	Language	Lexico-grammar
		Sentence
		Text
Regulative register	Task management	
	Behaviour management	

Figure 2. Coding scheme for analysing instructional practices in the observed lessons

Third, the assessment practices were analysed. As aforementioned, this paper only reports the analyses of written formative assessments, including worksheets and workbook exercises given by the teachers as in-class practice and homework. The questions of these written assessment tasks were coded according to a framework which consists of the same cognitive and linguistic levels as those listed under "instructional register" in Figure 2. Such a framework has been used to analysed questions in CLIL assessments in previous studies (Lo & Fung, 2020) and serves to gauge the different levels of cognitive and linguistic demands that assessment questions may impose on students. From the collected sample scripts, students' performance and teachers' grading practices shown in the marked scripts were also analysed.

In short, the objectives, instructional practices and assessment practices were analysed systematically using compatible frameworks. These frameworks had been tried out by three team members to analyse three lessons at the beginning of the data analysis phase. Modifications to the frameworks were then made and the coding procedures were clarified among team members, before the frameworks were applied to other observed lessons and assessment data. During the analysis procedure, the research team discussed whenever there was any uncertainty. Such analyses, together with teacher and student interview data, allow the researchers to explore and discuss the alignment among the three important components.

4. Results

4.1 Objectives

In the lessons observed, it seemed to be a normal practice for both teachers to articulate their lesson objectives to their students at the beginning of the lessons. For example, right at the beginning of the first lesson observed, Miss A told the students, *"For this lesson I want to tell you three things. Number one, we take some revision on the property of enzyme, okay? I hope you still remember what we talked about enzyme. Number two, I also want you to think about why we need to study enzyme, how they are important to our lives …"* Similarly, in her first lesson observed, Miss B also informed the students, *"For the coming topic, we are going to learn about common acids and alkalis … we are going to learn what are the acids commonly found in our daily life and also in the laboratory. Okay? And then their properties and what precaution we are going to take when we are handing these chemicals. And how to test the acidity and alkalinity."* Most of these articulated objectives tended to focus mainly on "content", particularly the knowledge or skills being covered. The only exception was found in Miss A's third lesson, where she explicitly highlighted a language-related objective, *"Now we need to learn two things today. The first thing is to describe the result from the graph …"* This can be regarded as both content and language objectives, as describing the results from a graph involves students' analytical ability (e.g., interpreting the graph and identifying different stages and the relationship between variables) and linguistic skills, particularly certain sentence patterns (e.g., *"As the temperature increases/decreases, the rate of reaction increases/decreases/remains unchanged"*).

4.2 Instruction

Using the devised coding framework (Figure 2), the lessons observed were analysed in terms of their orientation to "content" and "language". The results are shown in Table 2.

Table 2 shows that for both teachers, the percentage of "regulative" register (calculated by dividing the number of words coded as "regulative" register by the total number of words in that particular lesson) appeared to be rather high, with the mean being 14% and 24% for Miss A and Miss B respectively. This was mainly due to the experiments conducted in the lessons, in which teachers needed to give a lot of task instructions and manage students' behaviours from time to time (e.g., to urge students to hurry up or keep quiet).

Turning to "instructional" register, "content-oriented" episodes constituted an average of 71% and 95% (out of the total number of words in instructional

Table 2. Distribution of content-oriented and language-oriented episodes

	Miss A				Miss B			
	Lesson 1	Lesson 2	Lesson 3	Lesson 4	Lesson 1	Lesson 2	Lesson 3	Lesson 4
(i) Regulative register	6.9%	26.4%	8.9%	13.2%	15.6%	27.6%	33.2%	18.3%
(ii) Instructional register	(93.1%)	(73.6%)	(91.1%)	(86.8%)	(84.4%)	(72.4%)	(66.8%)	(81.7%)
– Content-oriented	74.4%	89.8%	32.8%	85.1%	91.5%	97.4%	96.7%	93.6%
– Language-oriented	25.6%	10.3%	67.2%	14.9%	8.5%	2.6%	3.3%	6.4%

Note. % out of the total no. of words in each lesson

register) for Miss A and Miss B respectively. This reveals that both teachers' lessons tended to be more "content-oriented". Comparatively speaking, there were more "language-oriented" episodes in Miss A's lessons. In particular, "language-oriented" episodes occupied 26% and 67% in lessons 1 and 3 of Miss A's lessons respectively. The exceptionally high percentage of language teaching in lesson 3 actually corresponded to Miss A's objectives for that lesson (i.e., to describe the result from the graph). These will be further discussed below.

When analysing the content-oriented episodes in detail (Table 3), we observed that there was some spread across the different cognitive levels in Miss A's lessons, with a mean of 40%, 54% and 6.8% for *recall*, *application* and *analysis* levels respectively. Most of the *recall* episodes were related to some factual information about the topic "enzymes" (e.g., the nature and functions of enzymes). The *application* episodes focused on the discussion about the application of enzymes, based on its nature and functions. Those *analysis* episodes took place when Miss A asked her students to critically evaluate the procedures of the experiment which might have affected the results. On the other hand, Miss B's lessons mainly focused on *recall* skills (87%), with some attention paid to *application* (13%) but none to *analysis*. Miss B focused on introducing to students the different kinds of acids and alkalis, as well as their functions, which making the lessons more oriented to *recall*. On some occasions, Miss B also discussed the application of acids and alkalis in daily life, and asked students to explain the results of the experiment. These occasions constituted the *application* episodes. Thus, Miss A's lessons appeared to be more cognitively challenging than Miss B's. Such differences regarding cognitive demands may be attributed to the different grade levels of the students (Grade 9 for Miss A vs Grade 8 for Miss B), the different topics involved ("Application of enzymes" for Miss A vs "Common acids and alkalis" for Miss B), and the general academic ability level of the students in the two schools. Despite such differences, both teachers were observed to provide conceptual scaffolding through verbal explanation, questioning, drawing on students' daily life experience, visual aids (e.g., pictures) and experiments. These are in line with previous studies (e.g., Lin & Wu, 2015; Pawan, 2008).

Table 3. Distribution of content-oriented episodes at different levels

Cognitive level	Miss A				Miss B			
	Lesson 1	Lesson 2	Lesson 3	Lesson 4	Lesson 1	Lesson 2	Lesson 3	Lesson 4
Recall	13.3%	59.3%	62.1%	24.0%	87.2%	87.0%	100%	75.7%
Application	82.8%	33.1%	37.9%	60.6%	12.8%	13.1%	0%	24.3%
Analysis	3.8%	7.7%	0%	15.4%	0%	0%	0%	0%

Note. % out of the total no. of words of the codes at cognitive level

Regarding the different levels of language-oriented teaching (Table 4), it was observed that for both teachers, the majority of language teaching episodes focused on teaching vocabulary or grammar. Moreover, these episodes were often "embedded" in content-oriented episodes. That is, when talking about the content knowledge, the teachers would temporarily shift students' attention to language by, for example, providing short definition for vocabulary or brief explanation of grammar features. Excerpt 1 serves as an example of such language-oriented episodes.

Table 4. Distribution of language-oriented episodes at different levels

Language level	Miss A				Miss B			
	Lesson 1	Lesson 2	Lesson 3	Lesson 4	Lesson 1	Lesson 2	Lesson 3	Lesson 4
Lexico-grammar	100%	100%	32.3%	47.5%	100%	100%	48.5%	100%
Sentence	0%	0%	12.9%	0%	0%	0%	51.6%	0%
Text	0%	0%	54.8%	52.5%	0%	0%	0%	0%

*% out of the total no. of words of the codes at language level

Excerpt 1. From Miss B's lesson 1 [08:22–08:43]

For the coming topic, we are going to learn about common acids and alkalis. You have to pay attention to the meaning of the word, acid. Acid 係酸, 得唔得? 酸. And alkali, 鹼. [*Acid is acid, okay? Acid. And alkali, alkali*] 咁呢兩個都係 noun 嚟嘅, 咁你會見呢眾數, 咁即係我地會學多過一種嘅酸同埋鹼啦 … [*So these two are noun. You can see the plural form. That means we will learn more than one type of acid and alkali…*]

In Excerpt 1, when talking about what students were going to learn in the coming topic, Miss B explained the key terms of the topic "acid" and "alkali" by providing L1 equivalents[1] (Lines 2–3). Then, when she mentioned that students were going to learn more than one type of acid and alkali, she drew students' attention to the part of speech and plural form of the key words (Lines 4–5). Indeed, such embedded vocabulary teaching episodes were highlighted by both teachers in the interviews, as they recognised the importance of teaching the subject-specific key words to the students. Such a practice has also been observed in other CLIL literature (Baecher et al., 2014), where CLIL teachers tended to focus more on vocabulary teaching when it comes to the language dimension.

Studying the trend of language-oriented teaching across lessons, it can be noticed that Miss A's lessons 3 and 4, and Miss B's lesson 3 consisted of more language teaching beyond the vocabulary level. However, for Miss B's lesson 3, the overall language-oriented teaching only accounted for 3% of the whole lesson (Table 2) and so it is more worth examining Miss A's lesson 3, in which over 65% of the lesson was language-oriented. It can be recalled that one key objective in Miss A's third lesson was "*to describe the results from a graph*". Hence, some language-oriented episodes in that particular lesson (and also some in lesson 4) focused on sentence patterns or text writing (i.e., how to write the paragraphs describing the results of the experiment). Excerpt 2 illustrates the first part of the teaching (with important parts underlined by the authors).

Excerpt 2. Miss A's lesson 3 [09:20–11:28]

… Then how would we describe this curve? "Describe the result" means describe this curve, you understand? Now, I will teach you. First of all, I will divide the graph into three parts. Why? Because these three parts, they will have their own special characteristics. Now, let's look at the first part. Now the first part, we talk about the temperature. The temperature will be high or low, compare with the other part?

---

1. It may be worth noting that Miss B switched to L1 (Chinese) from time to time in her lessons, while Miss A used English throughout her lessons. However, the use of language (L1 vs L2) is not the focus of this paper, and it does not affect the analysis (e.g., when the teacher used L1 to explain a grammar feature, it would be regarded as a "language-oriented" instructional episode).

Low.
Low, understand? So, how do we start with? <u>For the part, we say, at low temperature</u>, so this part is considered to be the temperature range is lower, understand? At low temperature, understand? <u>Now, so, as temperature increases [pause; writing on the blackboard], what happens to the rate of reaction? Now, I want you to learn this sentence pattern</u>. As temperature increases, we are looking at the rate, remember? What happens to the rate? You tell me for this part, part number one, how will you describe this? <u>Using similar words, we have increase, we have decrease, is that right? We have remain unchanged, understand?</u> Okay? Now which word will you choose for this part, part number one?
Increase.
<u>As temperature increases, so the rate of reaction,</u>
Increase
<u>Yes, also increases</u>. Is that right? Now, so it will call "describe part number one". <u>And then, part number two</u> …

As stated in Lines 1–2, Miss A was trying to teach the students how to describe the results (of an experiment) through describing the curve. She first divided the curve into three parts (Lines 2–3), and Excerpt 2 illustrates how she taught the first part and the sentence pattern "*As temperature increases, the rate of reaction also increases*". During the process, Miss A guided the students to analyse the trend (e.g., the temperature is low and it increases; what happens to the rate of reaction) and more importantly, how to describe what they observed. Before constructing the target sentence, Miss A elicited some key words from the students such as "*low*" (Line 7), "*increase*" (Lines 17 and 19) and introduced some key phrases including "*at low temperature*" (Lines 8–9) and "*as temperature increases*" (Line 10). After talking about the first part, she went on to the second and third part, which then constituted the short text that students were expected to write in their assignment (see more details below). In this way, Miss A provided explicit language scaffolding for the students to address a particular type of question (in this case, "describe and explain" question). In the post-observation interview, Miss A called such a practice as "worked-example strategy", which involved step-by-step illustration of the model answer to science problems in class. She adopted such a strategy in order to help students tackle assessment questions with appropriate language (e.g., key words, phrases, expressions). These may reflect Miss A's awareness of the important role played by language in assessments.

While such explicit language instruction (especially those beyond the "lexicogrammar" level) was absent in Miss B's lessons, in the post-observation interview, Miss B did share a few strategies that she would adopt with regard to language teaching, including teaching pronunciation of key words, explicit instructions on sentence patterns (e.g., compare and contrast), reading the textbook together with students, and repeating key concepts and phrases. However, due to time constraints (the observed lessons were scheduled towards the end of the semester), Miss B did not apply these strategies during classroom observations.

4.3 Assessment practices

Miss A

Miss A's formative assessment task was a take-home written assignment, which included one graph drawing question (to illustrate the results of an experiment) and three discussion questions based on the experiment. All these discussion questions require *application* skills. Two questions required students to read the questions presented in sentences, and to express their answers in sentences. The remaining one asked students to produce a piece of short text to explain the results of the experiment. This last question largely summarised what Miss A did in the four observed lessons. Hence, from the sample scripts collected, we analysed students' answers and Miss A's marking practices in detail.

Miss A awarded 10 marks for this question, and all students sampled performed quite well, getting 8 to 10 marks. Such good results could be attributed to Miss A's explicit language scaffolding in lessons 3 and 4 (part of it being illustrated in Except 2), which helped students to formulate their answers. In all the scripts collected, the students could address the question with the text structure and sentence patterns that Miss A talked about in the observed lessons. For example, in Excerpt 3, the student clearly separated her answer in three parts (paragraphs), each focusing on one part of the curve. She could also describe the curve with sentences like *"From 0º to 40ºC, as temperature increases, rate of reaction increases"*, which were illustrated by Miss A in the observed lessons. Hence, it is argued that Miss A's instruction in the lessons did help scaffold students to express their ideas more effectively and systematically.

Examining Miss A's marking practices, the researchers noticed that she seemed to focus more on the content, as she put ticks next to some key words/phrases in the scripts and then the total mark awarded corresponded to the number of ticks given. She would also give a general comment such as *"Very good"* and *"Your answer is accurate,"* but *"accurate"* here probably referred to accuracy of content. There were not many comments on students' language errors, probably because most students could produce rather well-formed and grammatical sentences. In the post-observation interview, Miss A also said that her grading rubrics were more content-oriented. For short questions, marks would not be deducted from language errors. Only if students hit all the key points but missed out on the language in long questions, one mark would be deducted. Such grading practices were also confirmed by Miss A's students in the student interviews. Some students admitted that as language errors were less important in summative assessment, they tended to focus more on the content (e.g., key words). This is perhaps an illustration of the backwash effect of assessment on students' learning behaviour.

Excerpt 3. A student's sample work from Miss A's class

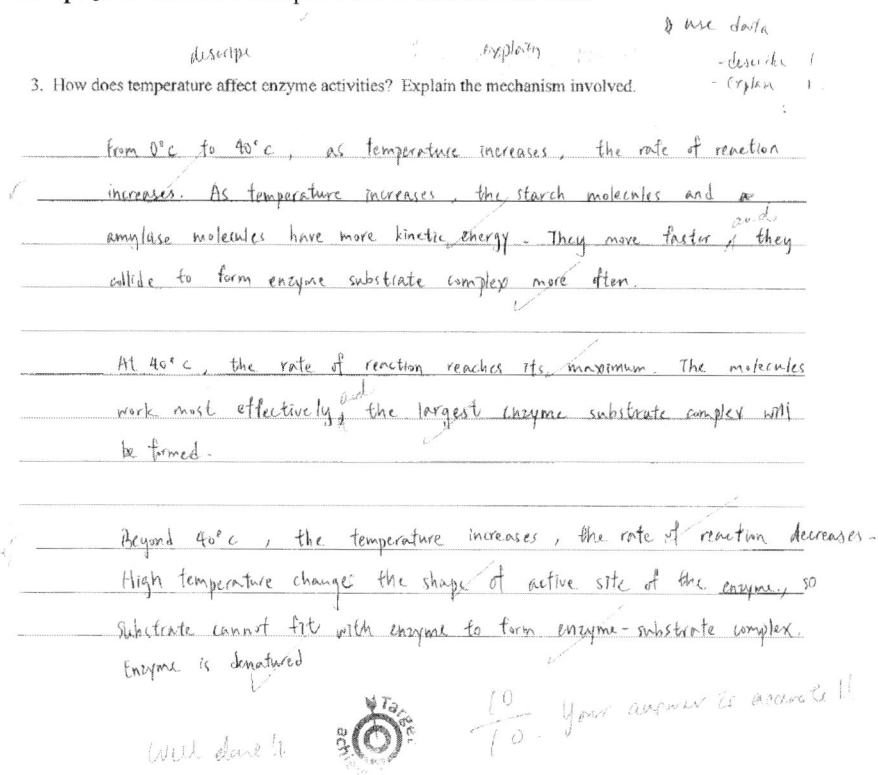

[Handwritten student work:]

describe | explain | § use data
— describe ✓
— explain ✓

3. How does temperature affect enzyme activities? Explain the mechanism involved.

From 0°C to 40°C, as temperature increases, the rate of reaction increases. As temperature increases, the starch molecules and amylase molecules have more kinetic energy. They move faster and they collide to form enzyme substrate complex more often.

At 40°C, the rate of reaction reaches its maximum. The molecules work most effectively and the largest enzyme substrate complex will be formed.

Beyond 40°C, the temperature increases, the rate of reaction decreases. High temperature change the shape of active site of the enzyme, so substrate cannot fit with enzyme to form enzyme-substrate complex. Enzyme is denatured.

Well done!! ♥Target achieved 10/10. Your answer is accurate!!

Miss B

Miss B's formative assessment task was a unit exercise in the workbook, which consisted of five True/False questions, five multiple-choice questions and three structured questions. When analysing these questions in detail, 40% required *recall* skills, 50% asked for *application* skills and the remaining 10% required *analysis* skills. Regarding linguistic demands, 77% of the questions did not require any language production, and around 10% of the questions asked students to produce vocabulary or sentences, and 13% asked students to write a short paragraph. Hence it seems that the cognitive and linguistic demands of the written formative assessment were not particularly high. Such relatively low demands actually aligned with the instructional foci of Miss B's lessons. These were probably due to the teachers' awareness of students' capacity, especially in relation to the potential language barrier. In the post-observation interview, Miss B mentioned that to help students understand the questions, structured questions tended to be presented in shorter sentences and did not expect students to write much. She admitted that these would allow students to express their content knowledge in English, but she was also aware that avoiding longer writing may prevent students from learning how to write more complete responses. However, in face of students' diversity, such a dilemma seems inevitable.

From the sample scripts collected, students' performance varied. As students were not required to produce much language, their different results were largely due to their understanding of the key concepts. For the only productive text-level question (i.e., "*Describe how a person can prepare a red cabbage extract*"), some students managed to write a rather coherent text, probably because they referred to the textbook. For example, one student wrote:

> He can chop the red cabbage leaves into small pieces and then put some small pieces of red cabbage leaves into a mortar and crush them with a pestle. Next put 30 cm^3 of distilled water into the mortar and stir the mixture with the pestle gently.

Such an answer could be regarded as rather complete, and hence was awarded the full mark. Some other students could only write incomplete sentences, such as "*Put some small pieces of red cabbage into a mortar and crush them with a pastle*", with some key words (e.g., pestle) being misspelt.

Miss B acknowledged her content-oriented marking practices in the interview, and such practices were also observed in the collected sample scripts. Similar to Miss A, she usually put ticks next to the target key words or points and then awarded marks. When there were misspellings or incomplete content, Miss B would use symbols to indicate them (e.g., circle the misspelt words; put "..." after the answer). Written feedback was rarely seen in the collected scripts.

5. Discussion and conclusions

This multi-case study seeks to examine to what extent CLIL content subject teachers' instructional practices align with the dual goal of CLIL programmes, and whether and how they provide scaffolding for students to cope with the dual challenges in CLIL assessments. Cross-case comparison reveals two major patterns of alignment in this study, which were demonstrated through the above illustration of the cases of Miss A and Miss B.

The first type of teachers, which was represented by Miss B and constituted the majority in this study, was heavily content-oriented. They focused more on the content knowledge when setting their lesson objectives, designing their instructional activities, assessment practices and marking rubrics. Owing to their content orientation, this group of teachers did not incorporate much explicit language scaffolding in their lessons. They mainly taught the key words of the topic, but seldom went beyond that to sentence or text level. Such findings corroborate the results of previous studies (e.g., Baecher et al., 2014; Tan, 2011). Although some of these teachers did demonstrate their language awareness during pre- or post-lesson observation chats and interviews, they were not able to put too much emphasis on explicit language instruction in their lessons owing to some contextual constraints such as the tight schedule before examination and their students' capacity. As illustrated in Miss B's case, owing to her awareness of the difficulties that students may encounter in language, her assessment tasks did not impose heavy linguistic demands. As a result, her students could still perform well in multiple-choice questions and structured questions which did not ask for complete sentences. Yet, Miss B admitted that some students were not learning the language along with the content knowledge. Hence, while there is still alignment among objectives, instruction and assessment for this group of teachers, such alignment may not correspond to the ideal "dual" goal of CLIL. As Gibbons (2015) argues, some teachers tend to lower the expectations or demands on students who may encounter language barriers, but it is crucial to provide a "high-challenge, high-support" environment to facilitate student learning and help them to succeed.

On the other hand, a couple of teachers were found to be more sensitive to both content and language teaching. This is the second type of teachers, represented by Miss A in this study. In addition to content coverage, some of their lessons had quite clear language objectives (e.g., teaching students how to describe a graph in a short paragraph). With these language objectives in mind, some of their lesson time and instructional activities were devoted to language scaffolding, during which students' attention was temporarily drawn to learning the academic language. Then, the teachers also expected students to demonstrate their language skills in the assessment tasks, which imposed higher productive language demands (e.g., writing coherent paragraphs). Thus, for this group of teachers, the alignment among objectives, instruction and assessment appears to pay more attention to the integration of content and language, thereby corresponding to the dual goal of CLIL, as represented in Figure 1.

Another point worth discussing is the potential "backwash effect" of assessment on teaching and learning behaviours observed in this study. For both groups of teachers, their assessment marking rubrics and practices were content-oriented. That is, they paid limited attention to "language", unless it concerns the spelling of subject-specific key words. Such a practice was recognised by the students, who admitted that they would focus more on the content of their answers and they knew that as long as they could get the key words correct, they would get the marks. Hence, it seems that teachers' assessment and marking practices have affected students' attitudes and efforts in the language aspect when learning content subjects in CLIL. This may not be conducive to the achievement of the dual goal of CLIL. However, the teachers admitted that they were actually following the marking rubrics of the high-stakes public examination. Miss A specially mentioned that if the examination authority did not make any changes in their marking rubrics, school teachers may not have the initiative to change their marking rubrics to give more recognition to the language dimension. Here, it may be worth examining the potential tension between the "official" learning and assessment objectives stipulated by the government and the "ideal" objectives implied in CLIL programmes. As highlighted previously, "(English) language teaching and learning" is not the objective in the official curriculum documents as well as assessment guidelines. This is likely to result in the content-oriented instructional and assessment practices of CLIL subject teachers (Tan, 2011). However, one should bear in mind that even though language learning may not be explicitly stated as one of the learning objectives, especially in those more content-driven programmes, the integral role played by "language" in CLIL cannot be overlooked. In other words, CLIL students still need to tackle the challenges in the linguistic aspect when they tackle content subject assessments (Lo & Fung, 2020; Shaw & Imam, 2013). Hence, it would be reasonable to expect CLIL subject teachers to incorporate some language scaffolding if their lessons are heavily content-oriented.

The findings of this study then yield some implications for CLIL practices and teacher education. First, considering the seemingly lacking of explicit language scaffolding in CLIL content subject lessons, together with the linguistic demands imposed by CLIL assessments, CLIL teachers could pay more attention to the language aspect in their lessons. They are encouraged to consider including some language objectives relevant to the content objectives in their lessons (Cammarata, 2016), and then incorporate some language scaffolding when delivering their lessons (Lin, 2016; Lyster, 2016). One useful strategy to incorporate language scaffolding specific for assessments, as identified in this study, is that teachers work on some assessment questions together with the students in lessons, so as to serve as demonstration or modelling. This may better prepare students to overcome the challenges in assessments.

Second, current professional development programmes for CLIL content subject teachers seem to focus more on teachers' academic language awareness and pedagogical practices. While these are definitely important, perhaps another component of these programmes could be on teachers' assessment awareness and assessment practices, so that CLIL teachers better understand how they can assess their students' learning progress and difficulties in both content and language dimensions, and how they can provide scaffolding for their students to overcome the challenges in CLIL assessments.

Third, it is interesting to note that while assessment questions impose both cognitive and linguistic demands on students, the marking rubrics emphasise the cognitive aspect only, resulting in some backwash on students' (and teachers') behaviours. Therefore, the examination authority or school administrators may consider putting more weight on the language aspect through, for example, adding effective communication marks or language bonus marks. In this way, teachers and students may have more incentive to put more efforts into language teaching and learning.

It is acknowledged that this multiple case study only focused on the practices of a few teachers in one educational context. In particular, the practices of the two teachers reported in this paper may have been affected by the characteristics of the students (e.g., general academic ability, English proficiency, age) and the topics being taught. Hence, it is not the intention of this paper to evaluate the practices of the participating teachers. Instead, it aims to explore any patterns of practices of CLIL teachers and yield some implications for pedagogy and teacher education, which can be further examined by research conducted in other CLIL educational contexts.

Acknowledgements

The study is supported by the Language Fund under Research and Development Projects 2015–16 of the Standing Committee on Language Education and Research (SCOLAR). We are grateful to the participating schools, teachers and students for their support. We also thank the anonymous reviewers for their suggestions.

References

Admiraal, W., Westhoff, G. J., & de Bot, K. (2006). Evaluation of bilingual secondary education in the Netherlands: Students' language proficiency in English. *Educational Research and Evaluation*, 121(1), 75–93. https://doi.org/10.1080/13803610500392160

Alderson, J. C. & Wall, D. (1993). Does washback exist? *Applied Linguistics*, 14(2), 115–129. https://doi.org/10.1093/applin/14.2.115

Baecher, L., Farnsworth, T., & Ediger, A. (2014). The challenges of planning language objectives in content-based ESL instruction. *Language Teaching Research*, 18(1), 118–136. https://doi.org/10.1177/1362168813505381

Cammarata, L. (2016). Foreign language education and the development of inquiry-driven language programs: Key challenges and curricular planning strategies. In L. Cammarata (Ed.), *Content-based foreign language teaching: Curriculum and pedagogy for developing advanced thinking and literacy skills* (pp. 123–143). New York, NY: Routledge. https://doi.org/10.4324/9780203850497

Cammarata, L., & Haley, C. (2018). Integrated content, language, and literacy instruction in a Canadian French immersion context: A professional development journey. *International Journal of Bilingual Education and Bilingualism*, 21(3), 332–348. https://doi.org/10.1080/13670050.2017.1386617

Cenoz, J., Genesee, F., & Gorter, D. (2014). Critical analysis of CLIL: Taking stock and looking forward. *Applied Linguistics*, 35(3), 243–262. https://doi.org/10.1093/applin/amt011

Christie, F. (2002). *Classroom discourse analysis: A functional perspective.* London: Continuum.

Coyle, D., Hood, P., & Marsh, D. (2010). *CLIL: Content and language integrated learning.* Cambridge: Cambridge University Press.

Earl, L. (2013). *Assessment as learning: Using classroom assessment to maximize student learning* (2nd ed.). Thousand Oaks, CA: Corwin Press.

Gablasova, D. (2014). Issues in the assessment of bilingually educated students: Expressing subject knowledge through L1 and L2. *Language Learning Journal*, 42(2), 151–164. https://doi.org/10.1080/09571736.2014.891396

Gibbons, P. (2015). *Scaffolding language, scaffolding learning: Teaching second language learners in the mainstream classroom* (2nd ed.). Portsmouth, NH: Heinemann.

Hönig, I. (2010). *Assessment in CLIL: Theoretical and empirical research.* Saarbrücken: VDM Verlag Dr. Müller.

Hughes, A. (2003). *Testing for language teachers* (2nd ed.). Cambridge: Cambridge University Press.

Kong, S., Hoare, P., & Chi, Y. P. (2011). Immersion education in China: Teachers' perspectives. *Frontiers of Education in China*, 6(1), 68–91. https://doi.org/10.1007/s11516-011-0122-6

Koopman, G. J., Skeet, J., & de Graaff, R. (2014). Exploring content teachers' knowledge of language pedagogy: A report on a small-scale research project in a Dutch CLIL context. *Language Learning Journal*, 42(2), 123–136. https://doi.org/10.1080/09571736.2014.889974

Krathwohl, D. R. (2002). A revision of Bloom's taxonomy: An overview. *Theory Into Practice*, 41(4), 212–218. https://doi.org/10.1207/s15430421tip4104_2

Li, D. C. S. (2017). *Multilingual Hong Kong: Languages, literacies and identities.* Cham: Springer. https://doi.org/10.1007/978-3-319-44195-5

Lin, A. M. Y. (2016). *Language across the curriculum: Theory and practice.* Dordrecht: Springer.

Lin, A. M. Y., & Wu, Y. (2015). 'May I speak Cantonese?' – Co-constructing a scientific proof in an EFL junior secondary science classroom. *International Journal of Bilingual Education and Bilingualism*, 18(3), 289–305. https://doi.org/10.1080/13670050.2014.988113

Llinares, A., Morton, T., & Whittaker, R. (2012). *The roles of language in CLIL.* Cambridge: Cambridge University Press.

Lo, Y. Y. (2014). Collaboration between L2 and content subject teachers in CBI: Contrasting beliefs and attitudes. *RELC Journal*, 45(2), 181–196. https://doi.org/10.1177/0033688214535054

Lo, Y. Y., & Fung, D. (2020). Assessment in CLIL: The interplay of cognitive and linguistic demands and their progression in secondary education. *International Journal of Bilingual Education and Bilingualism*, 23(10), 1192–1210. https://doi.org/10.1080/13670050.2018.1436519

Lo, Y.Y., & Jeong, H. (2018). Impact of genre-based pedagogy on students' academic literacy development in Content and Language Integrated Learning (CLIL). *Linguistics and Education, 47*, 36–46. https://doi.org/10.1016/j.linged.2018.08.001

Lyster, R. (2007). *Learning and teaching languages through content: A counterbalanced approach.* Amsterdam: John Benjamins. https://doi.org/10.1075/lllt.18

Lyster, R. (2016). *Vers une approche intégrée en immersion.* Montréal: Les Éditions CEC.

Massler, U., Stotz, D., & Queisser, C. (2014). Assessment instruments for primary CLIL: The conceptualisation and evaluation of test tasks. *Language Learning Journal, 42*(2), 137–150. https://doi.org/10.1080/09571736.2014.891371

Morton, T., & Jakonen, T. (2016). Integration of language and content through languaging in CLIL classroom interaction: a conversation analysis perspective. In T. Nikula, E. Dafouz, P. Moore, & U. Smit (Eds.), *Conceptualising integration in CLIL and multilingual education* (pp. 171–188). Bristol: Multilingual Matters. https://doi.org/10.21832/9781783096145-011

Morton, T., & Llinares, A. (2017). Content and Language Integrated Learning: Type of programme or pedagogical model? In A. Llinares & T. Morton (Eds.), *Applied linguistics perspectives on CLIL* (pp. 1–16). Amsterdam: John Benjamins. https://doi.org/10.1075/lllt.47.01mor

Navés, T. (2011). How promising are the results of integrating content and language for EFL writing and overall EFL proficiency? In Y. Ruiz de Zarobe, J.M. Sierra, F. Gallardo del Puerto (Eds.), *Content and foreign language integrated learning* (pp. 103–128). Bern: Peter Lang.

Orlich, D.C., Harder, R.J., Callahan, R.C., Trevisan, M.S., & Brown, A.H. (2013). *Teaching strategies: A guide to effective instruction* (10th ed.). Belmont, CA: Wadsworth Cengage Learning.

Pawan, F. (2008). Content-area teachers and scaffolded instruction for English language learners. *Teaching and Teacher Education, 24*(6), 1450–1462. https://doi.org/10.1016/j.tate.2008.02.003

Pérez-Cañado, M.L. (2016): Teacher training needs for bilingual education: In-service teacher perceptions. *International Journal of Bilingual Education and Bilingualism, 19*(3), 266–295. https://doi.org/10.1080/13670050.2014.980778

Reierstam, H. (2015). Assessing language or content? A comparative study of the assessment practices in three Swedish upper secondary CLIL schools (Unpublished doctoral dissertation). University of Gothenburg Retrieved from <http://hdl.handle.net/2077/40701>

Shaw, S., & Imam, H. (2013). Assessment of international students through the medium of English: Ensuring validity and fairness in content-based examinations. *Language Assessment Quarterly, 10*(4), 452–475. https://doi.org/10.1080/15434303.2013.866117

Tan, M. (2011). Mathematics and science teachers' beliefs and practices regarding the teaching of language in content learning. *Language Teaching Research, 15*(3), 325–342. https://doi.org/10.1177/1362168811401153

Wood, D.J., Bruner, J.S., & Ross, G. (1976). The role of tutoring in problem solving. *Journal of Child Psychiatry and Psychology, 17*(2), 89–100. https://doi.org/10.1111/j.1469-7610.1976.tb00381.x

Yin, R.K. (2009). *Case study research: Design and methods* (4th ed.). Thousand Oaks, CA: Sage.

Appendix 1. Lesson observation protocol

School & Teacher: _____ Topic/Curriculum Unit: _____
Subject: _____ Student Level: _____

Procedures/stages	Observer's notes		
	Lesson objectives	Instructional activities	Assessments
– What is being done? – Any key activities/ procedure(s)/step(s)?	– Cross-check with the teacher before class begins. – Content-based? Language-based? A mixture/balance of the two?	Things to pay attention to: – Pedagogy (e.g. How does the teacher help students understand the concepts/ language? How does the teacher help students answer questions?) – Content-based? Language-based? A mixture/balance of the two?	Things to pay attention to: – Formative assessment strategies (e.g. Are the questioning techniques related to vocabulary/ grammar/reading/ writing?) – Worksheets/ homework (e.g. What do they focus on?) – Summary assessment strategies (if any)

p. ____ of ____

Appendix 2. Examples of different registers when analysing the observed lessons

Category 1. *Instructional register, which focuses on knowledge delivery and discussion*

Category 1a. *Instructional register focusing on "content"*

Excerpt A. So, anybody remember what is the nature of enzyme?
Protein.
Which means enzymes – they are made up of what kind of substance?
Protein
Really, protein? Do you know what is protein? Anybody can tell me? What ____?
Amino acid.
Yes, what are the basic units? Amino acid – you still remember, right? …

Scaffolding for cognitive and linguistic challenges in CLIL science assessments 167

In this episode, Miss A focused on discussing the content (i.e., the nature of enzymes) with the students.

Category 1b. *Instructional register focusing on "language"*

Excerpt B. Indicator comes from the word indicate, 指示 [*indicate*]. 指示啲咩呢? [*What does it indicate?*] Indicator 就叫指示劑啦 [*Indicator is called indicator*]. Indicate 係個verb嚟, 即係指示 [*Indicate is the verb, which means indicate*]

In this episode, Miss B focused on the meaning and parts of speech of the key words, "indicate" and "indicator". Hence, it is coded as an instructional register focusing on "language".

Category 2. *Regulative register, which aims at managing classroom tasks and students' behaviours*

> Now, first of all, I want you to take out a reading exercise. I gave you a worksheet of enzyme – "What are enzymes" – last time, is that right, girls?
> Yes.
> Did you do some reading at home?
> Yes.
> Yes? Really? Okay.

In this episode, Miss A asked the students to take out a worksheet. The aim of this episode is to manage students' behaviours and is not directly related to instruction of lesson content.

The role of language in scaffolding content & language integration in CLIL science classrooms

Kok-Sing Tang
Curtin University

This chapter synthesizes the contributions from the authors in this edited volume by addressing two overarching questions. First, what is the role of language in mediating science teaching and learning in a CLIL science classroom? Second, to what extent can content and language be integrated or separated in CLIL instruction and assessment? In addressing the first question, I distil three major perspectives of how the authors conceive the role of language as a scaffolding tool. These roles are: (a) providing the discursive means and structure for classroom interaction to occur, (b) enabling students' construction of knowledge through cognitive and/or linguistic processes, and (c) providing the semantic relationships for science meaning-making. These three perspectives roughly correspond to the discursive, cognitive-linguistic, and semiotic roles of language respectively. In addition, two other roles – epistemic and affective, though not emphasized in this volume, are also discussed. In addressing the second question, I raise a dilemma concerning the integration of content and language. While there are clear political and theoretical arguments calling for an inseparable integration, there is also a common practice to separate content and language as distinct entities for various pedagogical and analytical purposes. In revolving this conundrum, I suggest a way forward is to consider the differences in the various roles of language (discursive/cognitive/linguistic vs. semiotic/epistemic/affective) or the levels of language involved (lexicogrammar vs. text/genre).

Keywords: classroom discourse, Content Language Integrated Learning (CLIL), role of language, scientific language, scaffolding

1. Introduction

The role of language is an area of special interest for many researchers and teachers both in language education and science education. The attention to language issues spans across a wide range of research interests, including classroom discourse, disciplinary literacy, reading- and writing-to-learn approaches, scientific language and genre, translanguaging, and more recently, the use of multimodal representations (e.g., diagrams, equations, gestures). Pertinent to this volume, the central focus revolves around the intersection of various home languages (e.g., English, Chinese, Japanese) and the language of science, within a political and pedagogical context of using a standardized medium of instruction. At the heart of this intersection is the drive toward a culturally relevant pedagogy that could cater to students' language proficiency and aspiration.

At the same time, the increasing prominence of Content and Language Integrated Learning (CLIL) in Europe and similar bilingual programs in other parts of the world has brought to attention the dual aspects of content-language teaching, which correspond to the goals of "learning language" and "learning through language" (Xu & Harfitt; this volume). Interesting, the idea of CLIL sparks an introspective reflection on what we typically mean by the terms "content" and "language" and how these two terms are related in the context of science teaching and learning. It also brings together two communities of people who have historically focused on one side of this duality – "content" for science educators and "language" for language educators (including L2 specialists). The timing of this volume is therefore excellent in calling for a greater convergence between these two communities to examine the complexity and nuances of content-language integration.

Against this background, Chapters 2 to 7 in this volume present novel classroom-based research that examines teachers' instructional practices and innovative pedagogies from a range of perspectives and foci, such as explicit language instruction or scaffold (An, Macaro, & Childs, this volume; Lo, Lui, & Wong, this volume), teacher language awareness (Xu & Harfitt, this volume), Concept + Language Mapping (He & Lin, this volume), translanguaging with *kanji* (Turner, this volume), and concept sketches (Ho, Wong, & Rappa, this volume). Collectively, the authors in this volume raised a number of insights that challenge and expand the central role of language in: (a) mediating science teaching and learning in general, and (b) its integration with content learning within CLIL classrooms more specifically.

In this commentary, I synthesize the contributions from the authors by exploring two overarching questions that span across their work. First, what is the role of language in mediating science teaching and learning in a CLIL science classroom? And second, to what extent can content and language be integrated or separated in CLIL instruction and assessment? The first question that guides my commentary is more broad and theoretical as I examine the authors' underlying (and sometimes implicit) theories and perspectives that inform the way they frame their research in their classroom intervention and/or discourse analysis of data. The second question is more contextual as I examine the pedagogical (and sometimes political) rationale and practices that either reinforce or constrain the integration of content and language in a CLIL context.

It should be mentioned that this commentary is written from the perspective of a science educator who has studied the use of language, literacy and discourse in more "mainstream" monolingual science classrooms, and has only some experiences in the CLIL research space. As such, the purpose of this commentary is not to provide critiques in the authors' theories and methodologies, but rather to highlight important ideas and issues that will foster further discussion and convergence among researchers working in multiple communities (e.g., applied linguistics, language education, bilingual education, science education) to move the field of CLIL forward.

2. Language in discursive, cognitive, linguistic, semiotic, epistemic & affective roles

What is the role of language in mediating the learning of science in a CLIL science classroom? The common answer to this question tends to rely on a neo-Vygotskyan notion of scaffolding, as explained in the introduction of this volume (Lo & Lin, this volume). Scaffolding is also a key construct that was foregrounded by several authors in their research design and analysis. For instance, "explicit language scaffold", "scaffolded interaction", and "designed vs. spontaneous scaffolding" are some constructs used by Lo et al. (this volume), Xu and Harfitt (this volume), and He and Lin (this volume) respectively. Although all the authors in this volume shared the same theoretical paradigm on scaffolding, there are some differences in the kinds of scaffolding they emphasized.

Reviewing across the six chapters, I distil three major aspects of how language was used as a scaffolding tool to support science teaching and learning in its various forms. The first aspect sees language as providing the discursive means and structure for classroom interaction to occur. The second aspect sees language as enabling students' construction of knowledge through cognitive and/or linguistic processes. The third aspect sees language as providing the semantic relationships for science meaning-making. These three aspects roughly correspond to the discursive, cognitive-linguistic, and semiotic roles of language respectively. In addition, I will discuss two other roles – epistemic and affective, which although were not emphasized in this volume, they are equally important in developing a holistic and balanced science education for all students.

2.1 Discursive role to scaffold classroom interaction

The first aspect of language as a discursive tool can be traced back to a long research tradition in classroom discourse. According to Cazden (1988, p. 3), "the study of classroom discourse is a kind of applied linguistics" to examine the language of teaching and learning. With roots in conversational analysis and interactional ethnography, researchers view language as a form of social action that is accomplished through contingent and turn-taking events, and consequently examine how language was used to structure classroom activity in ways that would be recognized and counted as teaching or learning. Historically, research in this area has moved from the classic IRF (initiate, response, feedback) pattern first documented by Sinclair and Coulthard (1975) decades ago to contemporary identification or prescription of effective "scaffolding strategies" or "talk moves" that engage students in meaningful dialogue. Such scaffolding strategies or moves apply to first language students, and would arguably have more importance for bilingual learners as they tend to be shut off from classroom conversation that involves a second or foreign language.

This discursive and performative aspect of language is best illustrated in this volume by Xu and Harfitt (this volume) and Ho et al. (this volume) in their analysis of scaffolding based on the classroom transcripts. To some extent, it is also observed in He and Lin's (this volume) analysis of the teachers' spontaneous scaffolding during classroom interactions. In Xu and Harfitt's study, they draw on Gibbons (2003) and Holton and Clarke (2006) to develop a set of "conceptual scaffolds" to understand how the teachers used language to co-construct knowledge with their students in the classroom. Such scaffolds include mediation, probing for expansion, translating, evoking discussion, encouraging self-scaffolding and withholding scaffolding. As for Ho et al.'s study, although they draw from a different framework by Chapin, O'Connor, & Anderson's (2013) focusing on questioning techniques, there are many overlaps with Xu and Harfitt's coding scheme, such as revoicing, probing, and clarifying.

In particular, Xu and Harfitt raised an interesting observation regarding the implicit and explicit nature of "scaffolded interaction", which I prefer to call discursive strategies. Discursive strategies are techniques that people use in a conversation (consciously or unconsciously) to achieve a particular purpose (Gumperz, 1982). IRF is a common discursive strategy used by teachers, many of whom are not even aware that they are using such technique to intuitively ask a narrow question with the aim of eliciting an expected answer. This is how many science teachers unknowingly or *implicitly* structure a conversation, particularly one that builds on a series of IRF patterns toward a specific content aligned with the official curriculum (Lemke, 1990). Xu and Harfitt suggest that some of the discursive strategies observed from the teachers were more conscious or *explicit* than other strategies. In my interpretation, a few strategies do indeed stand out in their explicitness to address the language challenges faced by the students. The first strategy is translating (episode 1, line 11), which interestingly is the only code that applies more specifically to bilingual students as compared to the rest of the codes that apply to all students. The second strategy is mediation (episode 2, line 5), which involves a comparison of terminologies used in scientific language as compared to vernacular language. The last one is withholding (episode 3), which shows a deliberate intention in holding back the answer.

This distinction between explicit and implicit scaffolding was also a major theme raised by other authors in this volume. However, instead of addressing the discursive role of language, what was being made explicit is quite different. In the next section, I will elaborate further, beginning with the cognitive-linguistic role of language.

2.2 Cognitive-Linguistic role to scaffold construction of knowledge

While the discursive role of language is necessary for organizing and accomplishing classroom interaction, it does not directly address the co-construction of knowledge that students are learning from the teachers. Xu and Harfitt (this volume) rightly expressed this particular role as follows: "Language is instrumental in the cognitive construction of knowledge and experience, not only because one disseminates knowledge by using a language, but also because language forms the way in which one construes an experience and internalises it as knowledge". In theorizing how language is used to scaffold the construction of knowledge, many authors in this volume draw from Vygotsky and Halliday to highlight the inseparable relationship between thought and language. Specifically, Halliday's systemic functional linguistics (SFL) provides a sophisticated framework to specify the linguistic structures that correspond to certain levels of knowledge construction. Due to this overlapping relation between thought and language, this is why I treat "cognitive-linguistic" as a unitary role of language, instead of separating them apart.

Lo et al. (this volume), An et al. (this volume), and Xu and Harfitt (this volume) provide good exemplars of this aspect of scaffolding that targets the cognitive and/or linguistic functions of language. In their chapters, their primary focus in making this cognitive-linguistic connection was to examine what kind of "content" and "language" was foregrounded in their case studies, thus determining how much content and language was integrated in CLIL. This content-language integration is an issue I will address in a later section, but for now, I will comment on their perspectives regarding the role of language as a cognitive-linguistic scaffolding tool.

A good place to start is the coding scheme developed by Lo et al., which categorizes a classroom video episode into, first of all, either an instructional or regulative register. Lo et al.'s emphasis is on the instructional register, which they further divided into content and language, with each aspect having three levels. Content, in this case, is a Bloom's taxonomy type of cognitive level, consisting of recall, application, and analysis. Language, based on Lin's (2016) adaptation from SFL, is divided into lexico-grammar, sentence, and text. Through the categorization and analysis of these episodes, Lo et al. brought out two important findings. In the first finding, the majority of language teaching episodes was found to be focused on vocabulary and "embedded" (i.e., implicit) in content-oriented episodes. The implicit nature of such vocabulary teaching episodes is similarly found in An et al.'s (this volume) analysis of language-focused episodes (LFEs) in their study, as well as many other studies in both CLIL and science education literature.

The second finding from Lo et al., which is more interesting, shows how a teacher (Miss A) provided a kind of *"explicit* language scaffolding" that went beyond vocabulary to sentence pattern and text writing. The example provided was the step-by-step instruction to describe an experimental result in the form of a graph. Linguistically, this instruction involves unpacking the noun phrases, nominalization, and conjunctive relations in the sentence pattern. Based on this analysis, Lo et al. argued that Miss A's instruction provided a better scaffold that enabled her students to understand and express their scientific ideas more effectively. Similarly, in An et al.'s study (this volume), although they did not see this kind of explicit instruction in their data, they also argued that such language structures are distinctive to the register of science, and should therefore be "systematically and explicitly instructed" for all science students, particularly more so for bilingual students.

2.3 Semiotic role to scaffold science meaning-making

The last aspect of scaffolding that I distil in this volume involves the semiotic role of language in providing the semantic relationships for science meaning-making. This perspective is influenced by the theory of social semiotics. With its roots from Halliday's meaning-oriented theory of language, social semiotics has developed into a general theory of how people make meanings using various semiotic modes, of which verbal language is just one mode among a range of multimodalities (Lemke, 1990). Social semiotics has often been used to analyze any text (broadly defined) by examining the semantic relationships made by every word, symbol, visual sketch, and body movement (e.g., Tang, 2011). Pedagogically, semantic relationships are also important as a scaffolding tool, and this is best illustrated by He and Lin (this volume).

He and Lin make use of Lemke's (1990) notion of thematic pattern, which is defined as "a pattern of semantic relationships", to expand a common strategy used in content-area teaching – concept mapping. As they rightly pointed out, although concept mapping provides a suitable "big picture" view of a science topic (consisting of multiple so-called "concepts"), it obstructs the fact that a concept itself is actually a shared and institutionalized thematic pattern that is repeated over and over again in multiple instances (e.g., teachers' talk, textbooks, written examinations). With this limitation of concept mapping, He and Lin developed what they called "concept + language mapping" in order to "emphasize the role of language in concept instruction in CLIL lessons". While thematic pattern has been used by researchers in text analysis, what is bold and unique in He and Lin's work is they have taken thematic pattern further to develop into a pedagogy. This will be useful for science teachers and students to become cognizant of the customary semantic relationships that are necessary to learn a particular scientific concept.

In Turner's (this volume) chapter, although she did not foreground semantic relationships or the meaning-making role of language, social semiotics provides an interesting interpretation to her key finding. Turner found from her study that the meanings of some scientific concepts were easier for the students to access in *kanji* than in English. From a social semiotics perspective, *kanji* is a unique semiotic system that was historically adopted from the Chinese writing scripts and later developed, along with the development of *hiragana* and *katakana*, to write parts of the Japanese language focusing on content words. As a character-based writing system, similar to written Chinese, the meaning of many words in *kanji* is realized through the semantic relationship of two or more characters. Taking an example from Turner's article, the meaning of "solid" in *kanji* is inferred from the semantic relationship of "hard", which is an ATTRIBUTE and "body", which is a MEDIUM (see Lemke, 1990 for a list semantic relationships). By providing information on the semantic relationships of a word from within the word itself, this was how *kanji* could be used as an additional resource for translanguaging between the students' L1 and L2. In this sense, the use of *kanji* as a scaffolding tool draws on the semiotic affordance of this written mode.

The example from *kanji* provides an interesting insight on how we learn vocabulary: the meaning of a new word can only be learned in relation to other existing words, or to be more exact, through their semantic relationships. This applies whether we learn a word from a dictionary or through a teacher's oral definition. Many CLIL teachers, as reported by An et al. and Lo et al. (this volume) usually stress on vocabulary instruction in their language-focused teaching at a word level. This can sometimes be problematic as bilingual students are led to rely on rote memorization and therefore mistake familiarity with scientific keywords for knowledge of the related concepts. Instead, learning a new scientific term (and its associated concept) often depends on learning the web of semantic relationships (or thematic pattern) that are used with the term. As such, the teaching of new technical vocabulary must not merely focus at a lexicogrammatical level (e.g., through repetition, synonym, paraphrasing), but also at a thematic-pattern level comprising several semantic relationships. This focus will involve a different instructional approach that makes explicit the underlying semantic relationships of a word, similar to what He and Lin (this volume) have done through their concept + language mapping approach.

Finally, Ho et al.'s (this volume) contribution to this volume is the foregrounding of the semiotic role of a visual representation. In their analysis, they provided two examples of how several features in the students' concept sketches served as a semiotic resource that complements the classroom talk. The first feature was the drawing of arrows to show directionality of H+ ions' movement, while the second feature was the drawing of enclosing shapes to show the "stuff inside" a stalked particle. Semantically, these two visual features represent transitivity relationship and part-whole (meronym) relationship (Kress & van Leeuwen, 1996), and they complement similar semantic relationships that were built through the oral discourse, for example in line 33 (H+ will *go out into* the...) and line 51–53 (What do you think is a stalked particle *made up of?*). Ho et al.'s primary focus in their chapter was to show how the verbal interaction (i.e., talk moves) guided the students' meaning-making of their diagrams. However, the reverse could also be argued and highlighted; that is, the classroom talk could not have occurred without the semantic relationships realized in the students' concept sketches. For instance, from line 44 and 46, there were many references to "*this* part" and "draw it *this* way" where presumably, the teacher was using gestures to point at several parts of the student's diagram. A more detailed multimodal analysis will therefore show the semiotic role of the drawing in scaffolding the classroom talk, and thus illustrate more generally how science meaning-making is mutually and symmetrically combined through a range of semiotic modes.

2.4 Epistemic and affective roles

Besides its discursive, cognitive, linguistic, and semiotic functions, language also has an epistemic and affective role that shape how students learn science language as well as learn science through its language. The language of science has a distinctive epistemic role that is integral to the construction of scientific knowledge (Shanahan, 2012). Some expressions of this language include the connection between claim and evidence, reasoning from theory, and the uncertainty of empirical finding. These expressions are embedded within certain communication patterns and structures in what many science educators call "scientific practices". The Next Generation Science Standards (NGSS) defines eight of these practices, which include investigation, using models, constructing explanation, engaging in argument and communicating information. From a language perspective, scientific practices roughly correspond to the notion of genre, which is conceptualized within SFL as the sociocultural context that shapes the production and interpretation of texts (Martin, 2007). In SFL, several researchers have also identified four major genres in science, namely experimental report, informational report, argument, and explanation (Halliday & Martin, 1993).

A number of chapters in this volume have findings that are relevant to the learning of scientific practices or genres. A good example is the language-oriented episode shown in Lo et al.'s (this volume) study where the teacher (Miss A) supported her students to "describe the results from a graph". In this episode, we could say that the class was partially involved in the genre of experimental report, and Miss A did well to unpack this particular genre for her bilingual students. However, Miss A focused more on the linguistic structure of the genre at a sentence and text level, which was understandable given that her students would need that kind of support. An epistemic focus, on the other hand, would manifest quite differently. For example, instead of learning the sentence pattern, the teacher would be asking students why they (or scientists in general) would choose this kind of language to describe their results or discussing the extent of how the graph (as a form of language) could represent the experimental results. For the genres of explanation and argument, an epistemic focus would question how a scientific explanation works based on an accepted theory, and likewise, how a scientific argument works based on empirical evidence (see Tang, 2016). Such epistemic emphasis is relevant not only in learning the content of science, but also the nature of science (NOS) itself.

Last but not least, the affective role of language is probably the most understudied in science education. This is likely because scientific language is often perceived to be factual, impersonal, and technical. Such interpersonal aspects of scientific language can present a learning barrier for children who are more comfortable with the personal and expressive forms of language used in everyday communication. The article by Turner (this volume) reminds me of how a second language like Japanese (or the children's L1) may provide an additional resource that could bridge the affective barrier of scientific language, which is taught through another language (e.g., English). In a study on English language learners (ELLs) conducted in Singapore with a colleague some years ago (see Wu, Mensah, & Tang, 2018), we observed that ELLs were translanguaging with their peers using L1 (Chinese) not in a conceptual manner to discuss scientific terminologies or meanings (e.g., cell nucleus), but more in a functional and social manner to express their emotions (e.g., disgusting over how plant cells look like under the microscope). This observation concurs with our knowledge and Turner's finding that bilingual learners tend to separate the use of their L1 from the official target language used to learn a content area like science, for a multitude of conceptual as well as social reasons. These social reasons warrant further investigation that considers the affective role of language in supporting bilingual learners in learning science.

3. Content & language as integrated or separate entities?

Having elaborated the discursive, cognitive, linguistic, semiotic, epistemic, and affective roles of language in scaffolding student learning of science, I will now discuss the second question: to what extent can content and language be integrated or separated in CLIL instruction and assessment? As this is not a simple question that can be resolved easily, I will only offer some observations based on my reading of this volume.

It is obvious that the central premise for promoting CLIL rests on the intertwined connection between content and language. This justification is based on not just a political convenience to support a country's language policies for a multilingual population, but also on a theoretical argument drawing upon the works of Vygotsky, Halliday, Lemke, and other theorists. Yet, in practice, it seems common to separate content and language as distinct entities for various pedagogical and analytical reasons. This somewhat paradoxical tension was partly raised by An et al. (this volume). While acknowledging the inextricable link between content and language within CLIL, they sought out to "establish whether it was actually possible" *for analytical purpose* to differentiate and identify "language focused episodes" that showed a teacher focusing on an aspect of language as opposed to "explaining about a scientific concept". A similar distinction can also be seen in Lo et al.'s (this volume) analysis where they had coded content and language as separate episodes in their video corpus.

There are of course many merits in separating content and language components for the purpose of analysis. One advantage is that the separation renders a complex phenomenon or case study into discrete categories or variables, which could then be further investigated. After this initial separation, the analyst's eventual goal is to put them back through some kind of meaningful comparison or connection. For example, while Lo et al. (this volume) coded content and language separately, they later showed how Miss A exemplified the objectives of CLIL by focusing more, as compared to Miss B, on *both* content and language aspects of science teaching. Pedagogically, it is also common to find CLIL lesson plans that have both content and language as distinctive learning outcomes. For instance, a content learning outcome could be "to state weight as the force of gravity and define it as the product of mass and acceleration due to gravity," while a language learning outcome could be to "use appropriate vocabulary related to describe forces and motion". Such distinction has the benefit of raising a teacher's awareness of achieving both content and language objectives at the same time in their CLIL lessons, even though both objectives are intertwined.

However, from other studies in this volume (e.g., He & Lin; Turner), it is also evident that what can be considered as content as opposed to language is sometimes difficult to differentiate. In these studies, content and language seem much more intertwined. In particular, I observe that specific to the discursive, cognitive, and linguistic roles of language, it is perhaps easier to differentiate content and language as two separate entities. By contrast, when discussing the semiotic, epistemic, and affective roles of language, the boundaries between content and language become more fuzzy and indistinguishable. In the semiotic role for instance, the understanding of "concepts", which is typically associated to content learning, was shown to be learned though the semantic relationships of language (realized through words, symbols, and other signs). In other words, learning the content is equivalent to learning the language, and these two aspects are actually two sides of the same coin.

Another way to think about these differences is through the various levels of language as used by Lo et al. and An et al. (this volume) in their analytical framework. At a lower level of language, such as lexico-grammar that deals with vocabulary, morpheme, spelling, punctuation, and noun phrases, it is perhaps easier to delineate content from the language aspect. For instance, if a student misspells a word or uses an improper grammar within a sentence, we can still easily infer the meaning (hence content) that he is trying to make based on the other words he has written in the sentence. In this case, we can ascribe the error as simply an expression issue and still able to discern the content separately. However, at a higher level of language (e.g., text-level) that deals with the thematic patterns and genres of science, it is not so easy to make the distinction. Thus, if a student does not have the language ability to put together the words and semantic relationships that form the thematic pattern of a concept, how else would he be able to demonstrate to others (and himself) an understanding of the content? This ability applies both during instruction when the student is learning a new concept through listening or reading as well as during assessment when the student has to demonstrate his understanding through oral discourse or a written test. Therefore, in scenarios that involve higher levels of language, it may not be feasible to separate content and language as distinct objectives or foci, for both analytical and pedagogical purposes.

4. Closing remarks

Chapters 2 to 7 in this volume have provided a number of thought provoking insights that expand our understanding of the roles of language, scaffolding, implicit vs. explicit instruction, and content-language integration. These contributions are important not only for researchers and teachers working in CLIL environments, but also more generally for science and other content-area educators who understood the importance of language. As I have mentioned earlier, although the language and science education research communities have come from different theoretical background and orientation, what draws both communities together is a common interest in understanding how the learning process in the science classroom is mediated by language, through its various discursive, cognitive, linguistic, semiotic, epistemic, and affective functions.

Based on this volume, it is perhaps understandable why the science classroom provides such a fascinating research site that draws researchers from both communities. Among the various disciplines, science is highly regarded as a "content-heavy" subject. Bernstein (1999) describes the natural sciences as having a hierarchical knowledge structure where the content knowledge is accumulative as opposed to a horizontal knowledge structure in the humanities. At the same time, we have also learned from years of previous research that the language of science (including the use of mathematics and other multimodal representations) is distinctively different from everyday communication. As such, the teaching and learning of science provide a favourable combination of content and language for researchers to explore and investigate. For CLIL researchers, the distinction between content and language in the context of science, as well as their integration in the classroom, therefore become a point of interest that warrants further investigation.

In this regard, as someone who works mostly within science education, I am delighted to see more researchers from the CLIL community coming to investigate this interesting space and sharing our intellectual resources to tackle an important educational problem. In closing, I thank all the authors in volume for being the pioneers in this field. As they are forging the integration of content and language within CLIL science classrooms, they are also simultaneously advancing a deeper convergence between the science education and language education communities at large.

References

Bernstein, B. (1999). Vertical and horizontal discourse: An essay. *British Journal of Sociology Education*, 20(2), 157–173. https://doi.org/10.1080/01425699995380

Cazden, C. B. (1988). *Classroom discourse: The language of teaching and learning* (1st ed.). Portsmouth, NH: Heinemann.

Chapin, S., O'Connor, C., & Anderson, N. (2013). Classroom discussions in Math: A teacher's guide for using talk moves to support the common core and more, Grades K-6: A Multimedia Professional Learning Resource (third edition). *Sausalito*, CA: Math Solutions Publications.

Gibbons, P. (2003). Mediating language learning: Teacher interactions with ESL students in a content-based classroom. *TESOL Quarterly*, 37(2), 247–273. https://doi.org/10.2307/3588504

Gumperz, J. J. (1982). *Discourse strategies*. Cambridge: Cambridge University Press. https://doi.org/10.1017/CBO9780511611834

Halliday, M. A. K., & Martin, J. R. (1993). *Writing science: Literacy and discursive power*. Pittsburgh, PA: University of Pittsburgh Press.

Holton, D., & Clarke, D. (2006). Scaffolding and metacognition. *International Journal of Mathematical Education in Science and Technology*, 37(2), 127–143. https://doi.org/10.1080/00207390500285818

Kress, G., & van Leeuwen, T. (1996). *Reading images: The grammar of visual design*. London: Routledge.

Lemke, J. L. (1990). *Talking science: Language, learning and values*: Norwood, NJ: Ablex.

Lin, A. M. Y. (2016). *Language across the curriculum: Theory and practice*. Dordrecht: Springer.

Martin, J. R. (2007). Genre, ideology and intertextuality: A systemic functional perspective. *Linguistics and the Human Sciences*, 2(2), 275. https://doi.org/10.1558/lhs.v2i2.275-298

Shanahan, M. C. (2012). Reading for evidence through hybrid adapted primary literature. In S. P. Norris (Ed.), *Reading for evidence and interpreting visualizations in mathematics and science education* (pp. 41–63). Rotterdam: SensePublishers. https://doi.org/10.1007/978-94-6091-924-4_3

Sinclair, J. M., & Coulthard, M. (1975). *Towards an analysis of discourse: The English used by teachers and pupils*. London: Oxford University Press.

Tang, K. S. (2011). Reassembling curricular concepts: A multimodal approach to the study of curriculum and instruction. *International Journal of Science and Mathematics Education*, 9, 109–135. https://doi.org/10.1007/s10763-010-9222-7

Tang, K. S. (2016). Constructing scientific explanations through premise–reasoning–outcome (PRO): An exploratory study to scaffold students in structuring written explanations. *International Journal of Science Education*, 38(9), 1415–1440. https://doi.org/10.1080/09500693.2016.1192309

Wu, S. J., Mensah, F. M., & Tang, K. S. (2018). The content-language tension for English language learners in two secondary science classrooms. In K. S. Tang & K. Danielsson (Eds.), *Global developments in literacy research for science education*. Cham: Springer. https://doi.org/10.1007/978-3-319-69197-8_8

Index

A
academic language   4–5, 146, 152
academic literacy   9, 35, 116, 146 *See also* language of science; scientific language
affective   44, 171, 172, 176–179
assessment: formative assessment   150–151, 152, 158–159
    summative assessment   150–151, 159

B
backwash effect   144, 147, 159, 161
bilingual teachers   18, 19, 36

C
clarification: clarification request   27, 71
    seek clarification   97, 99, 100, 101, 113
classroom discourse   70, 172
classroom talk   72, 76, 80, 86, 87, 88–89, 92, 102, 104–105, 107, 108, 176
co-construction: co-construction of knowledge   71, 173
    co-construction of concept meaning   127
    knowledge co-construction   89
Cognitive Discourse Functions (CDFs)   45
collaborative learning   128, 130
communicative approach   89, 120, 128
communicative methodology   18
communicative language classrooms   33
communicative language use   44, 55
Concept + Language Mapping (CLM)   10, 118–120, 129, 137, 175
CLM approach   118–119, 135–137, 138
CLM pedagogy   119, 134, 135, 136–137
concept mapping   116–117, 118, 175
concept sketch   10, 87, 88, 91, 92, 105, 107, 176
conceptual scaffolding   8, 75–76, 80, 155
content objectives   146, 162
content-oriented episodes   154, 155, 174
co-teaching   49, 57

D
dialogic   88, 89, 120, 129–130
discursive strategies   172–173

E
elicitation   93, 95, 96, 100, 102
English Medium Instruction (EMI)   2, 3, 18, 19, 64, 135, 136, 144, 148–149
epistemic   176–177
explicit instruction   29
explicit instruction of language   5, 32, 33, 38
explicit language instruction   157, 161, 170
explicit language scaffolding   157, 158, 161, 162
explicit teaching   24, 35

G
general academic vocabulary   21, 34
genres   4, 177, 179
grammatical metaphor   20

I
Initiation-Response-Feedback (IRF)   70, 76, 172, 173

L
language objectives   145, 146, 161, 162
language of science   4, 20, 85, 106, 136, 170, 177, 180
language-aware (content) teacher   64, 66, 70
language-focused episode (LFE)   9, 21–22, 32–35
language-oriented episode   152, 155, 156, 177

M
meaning-making   86–87, 106, 107, 174–176
mediate   79, 88, 106
mediation   71, 75, 78, 80, 173
modes   86–87, 89, 103, 105, 175
monolingual teachers   19, 20, 32, 33, 34–36
multimodal: multimodal literacy   85, 89
    multimodal representation   170, 180
    multimodality   8, 89, 103
    multimodalities   175
    Multimodalities-Extextualisation-Cycle (MEC)   8

N
negotiation of meaning   19, 28, 37
nominalization   20, 35 *See also* grammatical metaphor
non-technical vocabulary   21, 25–27, 28, 33–34

Index

P
pedagogical content knowledge 47, 66
pedagogical language knowledge 47
probing 75, 76, 80, 103, 105, 172

R
reformulate 71, 76, 105
reformulation 80, 102
registers 4–5, 75
repetition with variation 119, 134–135

S
scaffolded interaction 67, 80, 171, 172
scaffolding: designed scaffolding 119–120, 128–129, 130, 135, 136, 137
   instructional scaffolding 148
   interactional scaffolding 105
   linguistic scaffolding 8
   spontaneous scaffolding 10, 120, 130–134, 135, 137, 172
scaffolding strategies 75–76, 80, 172
scientific language 46, 73, 170, 177
semantic patterns 117
semantic relationships 117, 130, 134, 137, 174–176, 179
semiotic resources 10, 86, 87
social constructivism 86
social constructivist 6–7, 71, 147
social semiotics 174–175
subject-specific terminology 146
subject-specific vocabulary 4, 126 *See also* technical vocabulary
Systemic Functional Linguistics 4, 35, 173

T
talk moves 91, 113, 172
teacher language awareness 10, 63, 65, 79–80
teacher questioning 88, 105, 106
technical vocabulary 20–21, 25, 29, 34, 176
thematic patterns 10, 116, 117–118, 119, 130–131, 133–134, 135–136
translanguaging 37, 78, 175, 178

V
visual representation 10, 87, 88, 105, 117
visuals 8, 9, 10